Automotive Computer Systems and Circuits

Edward J. Ralbovsky
Automotive Technology Department,
State University of New York at Canton

Francis N. Treichler
Ford Motor Company
North Tonawanda, New York

PRENTICE HALL
Englewood Cliffs, New Jersey 07632

Library of Congress Cataloging-in-Publication data

Ralbovsky, Edward
Automotive computer systems and circuits/Edward J. Ralbovsky
Francis N. Treichler
p. cm.
Includes index.
ISBN 0-13-054230-X
1. Automobiles—Electronic equipment. 2. Microprocessors.
I. Treichler, Francis N., (*date*). II. Title.
TL272.5.R35 1988
629.2'549—dc19

Editorial and production supervision
and interior design: **PATRICK WALSH**
Cover Design: **BEN SANTORA**
Manufacturing buyer: **PETER HAVENS**

Printed in the United States of America

10 9 8 7 6 5 4 3 2 1

ISBN 0-13-054230-X

Prentice-Hall International (UK) Limited, *London*
Prentice-Hall of Australia Pty. Limited, *Sydney*
Prentice-Hall Canada, Inc., *Toronto*
Prentice-Hall Hispanoamericana, S.A., *Mexico*
Prentice Hall of India Private Limited, *New Delhi*
Simon & Schuster of Asia Pte. Ltd., *Singapore*
Editora Prentice-Hall do Brasil, Ltda., *Rio de Janeiro*

Contents

Preface

The modern automobile is truly a revolutionary piece of equipment when you consider that in the first 70 years of its existence very little about it has changed. Appearances have been altered somewhat and reliability has improved, but generally construction techniques have stayed about the same. Mechanical operation of most systems and components could be figured out rather easily and became common knowledge. Until ten years ago, the only electronic device in the automobile was the radio, and that was crude by today's standards. Since then we have seen the use of new construction techniques and materials, new types of drive train components, more kinds of comfort options, and electronic controls for almost every automotive system.

For the people servicing these new vehicles (and the people who train service people), the new materials and mechanical devices were not too difficult to master, since they were a natural evolution of previous systems. But electronic controls have proved to be another matter altogether. Electrical systems have always seemed difficult for most auto mechanics (and teachers). Electronic control operations are perceived as more mysterious and foreboding, and the many attempts at explanation have simply added to the confusion.

The authors wrote this book out of the need to provide answers to students and technicians who were asking, "How do these electronic systems work?" The purpose of this book is to provide clear, logical explanations of sufficient depth to enable student and technicians to comprehend what is happening. We believe this will help them to properly (and profitably) service electronically controlled systems.

Chapter 1 provides electrical information that is used throughout the text and must be thoroughly understood before proceeding to the other chapters. It was written without knowing the electrical background of the reader. Only electrical information essential for understanding the following chapters is provided.

Chapter 2 divides automotive computer systems into three basic sections: Inputs, the Computer, and Outputs. Each section is subdivided into the individual circuits and components that are explained. The rationale for this is that manufacturers will use different combinations of inputs and outputs with the computer. So when the students encounter the different systems, they will see what circuits are used and will know how they work.

Chapter 3 discusses the purpose of engine computer systems and the different operating modes. Computer control of carburetors, central fuel injection, ported fuel injection systems, and testing procedures are covered.

Chapter 4 covers other automotive computer systems such as electronic instrumentation, antilock brakes, air suspension, and cruise control.

References to specific car lines and manufacturers have been avoided, since all manufacturers use similar systems in various applications. To go through each system by brand is unnecessary, redundant, boring, and quickly becomes outdated. We believe that once you understand how the inputs, outputs, and computers operate, that you will be able to apply this knowledge to the many combinations of computer systems being used. You will be able to identify what circuits and systems are being used and you will know how they operate. We have also avoided reprinting manufacturers diagnostic procedures since they are applicable to specific models at specific years. To print a diagnostic routine and have the student use a routine that is out of date or changed by the manufacturer invites confusion and frustration. We strongly recommend the use of service manuals when working on a vehicle. It is assumed that you have a genuine interest in obtaining up-to-date, factual information about this facet of automobile servicing.

It has been our pleasure to research and present this information to you.

ACKNOWLEDGMENTS

The authors are grateful to those who have helped us with the development of this book. We especially wish to thank the following:

Ford Motor Company

General Motors Corporation

Our friends and colleagues at the State University of New York, College of Technology, at Canton, New York.

Our reviewers, William LaPierre and Bill Johnson

Heidi and Ruthie

Thank you,

Ed & Fran

Chapter 1

Electrical Laws You Need to Know to Understand Automotive Computer Circuits

OBJECTIVES

After studying this chapter, the student will be able to

- define amperes, volts, and ohms
- know the common division of units
- state Ohm's Law
- state series and parallel circuit characteristics
- perform series and parallel circuit calculations
- describe the operation of a simple transformer
- define capacitance and state three factors governing capacitance
- know the differences between an analog and digital signals
- describe the function of the diode
- describe operation of the transistor

§ 1-1 INTRODUCTION

All matter is made of atoms. Atoms are made of protons, neutrons, and electrons, Figure 1-1. Electrons are the particles that make up electrical current. When electrons move from atom to atom you have electrical current. Materials that allow for easy movement of the electrons are called conductors. Materials opposing electron movement are called insulators, and materials that can take on the characteristics of either a conductor or an insulator are called semi-conductors.

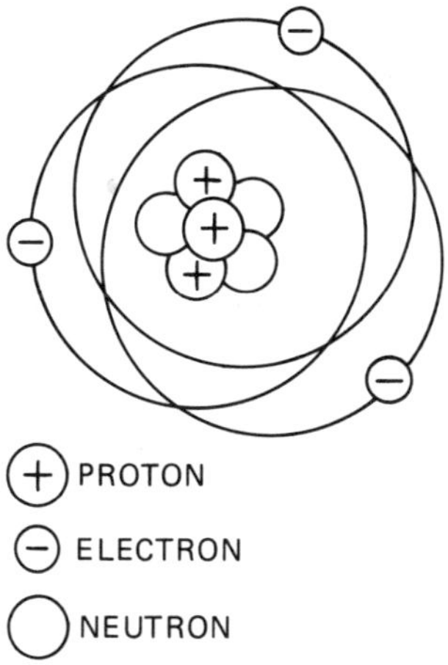

Figure 1-1 An atom.

1-1-1 Current

As stated, electrical current is the movement of electrons. This flow of electrons is measured in units called amperes (or amps). One ampere (a) equals 6.25×10^{18} electrons that have passed a given point in one second. An ampere represents the rate at which electrons flow past a given point. The more electrons that flow by a given point in one second, the higher the amperage or current flow. The letter I is a symbol used to represent the ampere; it literally stands for intensity, and is used when doing calculations.

1-1-2 Voltage

Voltage can be thought of as the force that moves electrons. In order to get current flow, an electrical pressure must be applied to the electrons. That electrical pressure is voltage. To increase current flow, a greater voltage (pressure) must be applied. The unit used to measure electrical pressure is the volt (v). When doing calculations, voltage is often represented by the letter E, which literally stands for electromotive force.

1-1-3 Resistance

When electrons move, there is always some resistance. Resistance is the opposition to current (electron flow). The unit for measuring electrical resistance is the ohm. One ohm will allow one ampere to flow when one volt is applied. The letter R is used to represent resistance and is used for calculations. Resistance values are indicated with the omega symbol (Ω).

1-1-4 Units of Measurement

There are four essential prefixes you need to know. They are: Mega (M), Kilo (k), Milli (m), and Micro (u). They literally mean:

M = Mega = 1,000,000 = one million
k = Kilo = 1,000 = one thousand
m = Milli = 1/1000 or .001 = one/one thousandth
u = Micro = 1/1,000,000 or .000001 = one/one millionth.

These prefixes are attached to amps, volts, and ohms. Here are some examples:

20 microamps = 20 ua = .000020a or 20/1,000,000a
40 millivolts = 40 mv = .040v or 40/1000v
20 kilovolts = 20 kv = 20,000v
40 megaohms = 40 MΩ = 40,000,000Ω

These four prefixes are commonly used when discussing automotive computer circuits.

§ 1-2 CURRENT FLOW AND DIRECTION

For electrons to flow, there must be a complete path with unequal charges at opposite ends. The pathway is called a circuit, and it consists of a power source and a conductor, Figure 1-2. The power source, in this case a battery, has more electrons at the negative terminal than at the positive terminal. The excess electrons at the negative terminal repel each other, and they are attracted to the positive terminal where there are far fewer electrons. The electrons will travel along the conductor so long as there is a difference of charge from the negative to the positive terminal. Current will stop flowing when both terminals have the same number of electrons and, therefore, each terminal has the same charge.

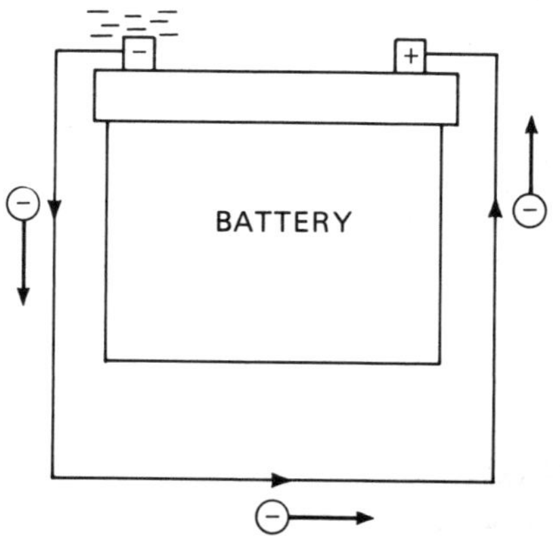

Figure 1-2 Electron flow

The electron flow from negative to positive is called the Electron Theory of Current Flow. The Electron Theory is used to help describe semiconductors such as diodes and transistors.

When tracing electrical circuits and describing circuit operation and troubleshooting, it is often more convenient to trace circuits from positive to negative. This book will trace circuits from positive to negative unless stated otherwise.

§ 1-3 OHM'S LAW

Ohm's law is the key to understanding electrical circuits. Ohm's law states: The current in a circuit is directly proportional to the applied voltage and inversely proportional to the resistance. Here are the implications of this law:

- When voltage increases or decreases, current will increase or decrease if resistance is fixed.
- When voltage is up—amperage (current) will be up if the resistance is constant.
- When voltage is down—amperage (current) will be down if the resistance is constant.
- When resistance increases or decreases, current decreases or increases if the voltage is fixed.
- When resistance is up—amperage (current) will be down if the voltage is constant.
- When resistance is down—amperage (current) will be up if the voltage is constant.

Mathematically, Ohm's law is expressed as follows:

$$E = I \times R$$

$$R = E/I$$

$$I = E/R$$

If you know any two values, you can find the third. An easy way to remember is to draw a circle with an E in the top half, and an I and R in the lower half, Figure 1-3. Cover the unknown value with your finger and you will obtain the formula. For example, to find resistance, cover the R symbol to obtain E/I.

Ohm's law will be applied in series circuits, parallel circuits, computer circuits, and troubleshooting.

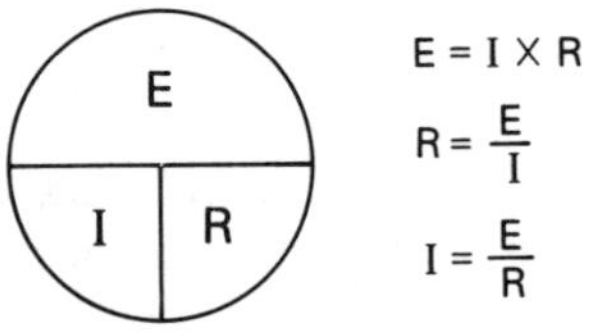

Figure 1-3 Ohm's law.

§ 1-4 SERIES CIRCUIT CHARACTERISTICS AND CALCULATIONS

The primary characteristic of a series circuit is that the electrons have only one path to travel. In the simple circuit, there is a power source, one current path, and one load, Figure 1-4. Electrons coming out of the battery have no choice but to follow the conductor, go through load R, and return to the battery.

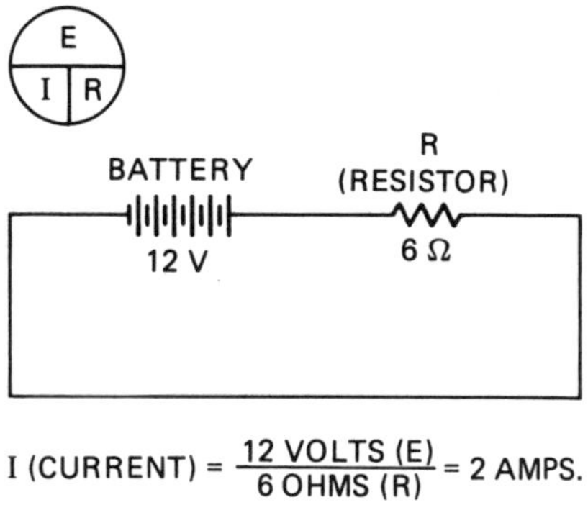

Figure 1-4 Simple circuit.

A true series circuit has two loads or more placed end to end with only one current path, Figure 1-5. There are three characteristics of simple and series circuits you will need to know.

- The sum of the individual resistances equals the total resistance (R).
- The amperage is the same at any point within the series circuit.
- The total voltage used up in a series circuit equals the applied (source) voltage (V). Voltage used up is referred to as voltage drop.

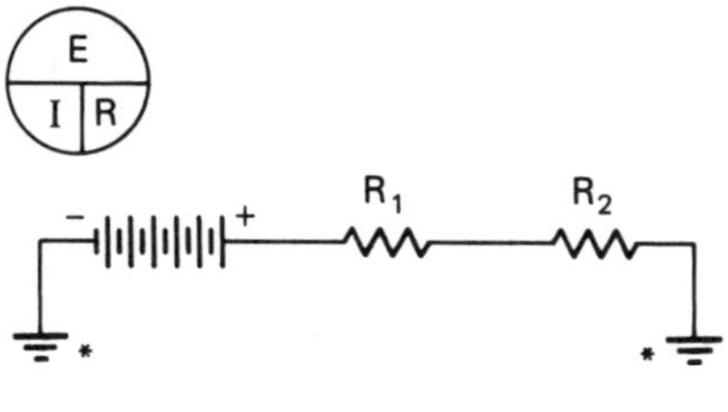

Figure 1-5 Series circuit. * Denotes a common connection to a conductor—commonly called a ground.

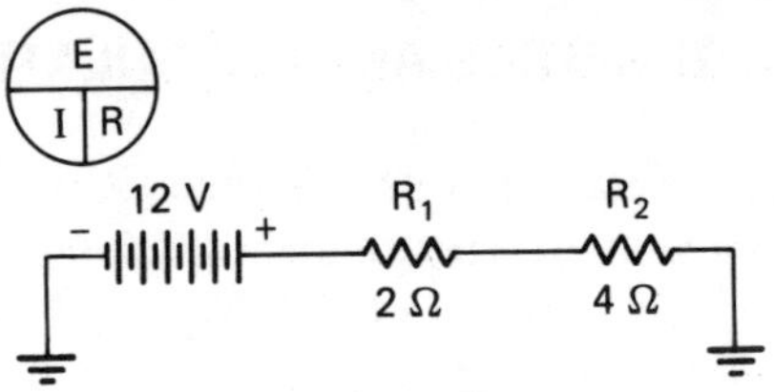

$$R_t = R_1 + R_2$$
$$R_t = 2\,\Omega + 4\,\Omega$$
$$R_t = 6\,\Omega$$

Figure 1-6 Series circuit with a total resistance of 6 ohms.

1-4-1 Resistance in a Series Circuit

The first rule, the sum of the individual resistances equals the total resistance, is expressed mathematically as: R1 + R2 = Rt. (If there are more than two resistors, they are simply added in.) Thus, if a series circuit has a 2-ohm resistor and a 4-ohm resistor, total resistance (Rt) would equal 6 ohms, Figure 1-6. What you need to remember is that to find total resistance in a series circuit you must add together all the individual resistances.

1-4-2 Amperage in a Series Circuit

Amperage in a series circuit is the same anywhere in the circuit. To find the amperage in a series circuit, divide the source voltage by the total resistance, Figure 1-7. As stated earlier, amperage is the number of electrons passing a point in one second. In the series circuit, with only one path to follow, the same number of electrons that enters the circuit must also leave the circuit. An ammeter placed anywhere in the circuit will always give the same reading.

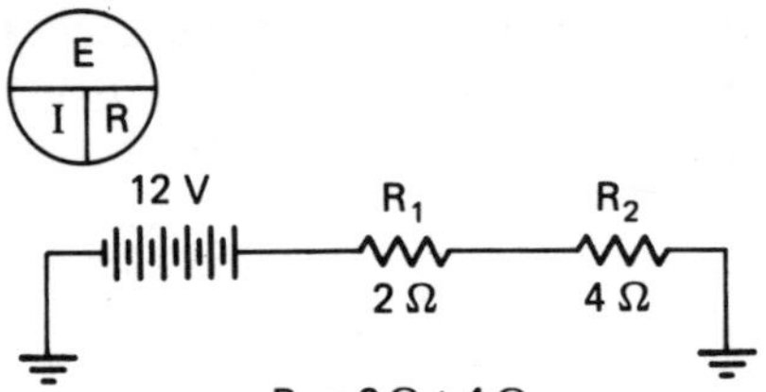

$$R_t = 2\,\Omega + 4\,\Omega$$
$$R_t = 6\,\Omega$$
$$I = E/R$$
$$I = 12\text{ V}/6\,\Omega$$
$$I = 2\text{ A}$$

Figure 1-7 Series circuit with amperage of 2A.

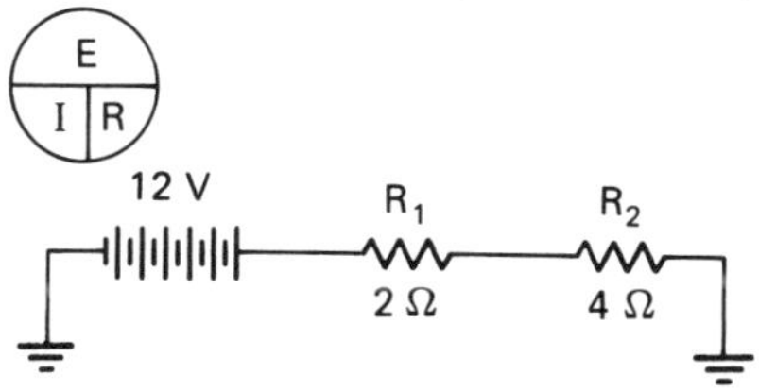

TO FIND THE VOLTAGE DROP FOR R_1

$E = I \times R$

$E = 2\,A \times 2\,\Omega$

$E = 4\,V$

TO FIND THE VOLTAGE DROP FOR R_2

$E = I \times R$

$E = 2\,A \times 4\,\Omega$

$E = 8\,V$

Figure 1-8 The calculated current for the circuit was 2 amperes. Since current is the same in all parts of a series circuit, 2A would be flowing through both R_1 *and* R_2.

1-4-3 Voltage/Voltage Drops in a Series Circuit

The second rule in a series circuit, the sum of the individual voltage drops equals the source (applied) voltage, is more complex but must be understood fully, for this rule is used extensively in computer circuits.

First, whenever current passes through a load, voltage decreases. The greater the resistance, the greater the voltage decrease. This decrease in voltage is called the voltage drop. The amount of voltage dropped is directly proportional to the resistance. To calculate the voltage drop, you need to use Ohm's law where you multiply the amperage times the resistance (Figure 1-8).

Note that R2 has greater resistance and, therefore, a higher voltage drop. Also, the four volts for R1 plus the eight volts for R2 equals twelve volts which is the source voltage. The sum of the voltage drops must equal source voltage. If they do not, a mistake was made in the calculations or a resistance source was left out.

The examples given so far deal with values you may already be familiar with. The following are examples using Ohm's law and series circuit characteristics, but with values that are typical in today's computer circuits (Figure 1-9).

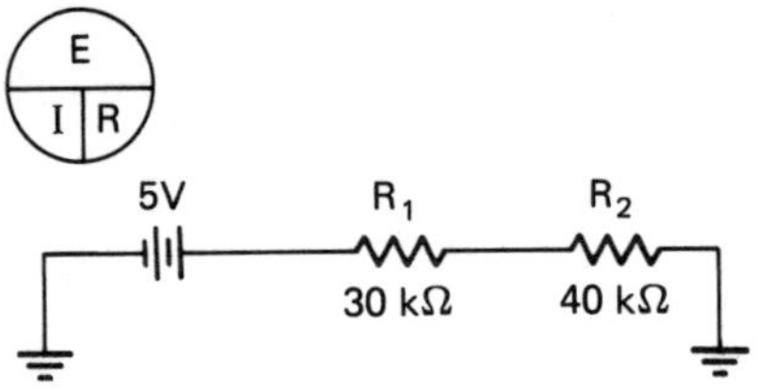

TO FIND $R_t = R_1 + R_2$

$R_t = 30{,}000\ \Omega + 40{,}000\ \Omega$

$R_t = 70\ \text{k}\Omega$

TO FIND $I = E/R$

$$I = \frac{5\ \text{V}}{70{,}000\ \Omega}$$

$I = 71\ \mu\text{A}$

TO FIND VOLTAGE DROP FOR R_1

$E = R_1 \times I$

$E = 30{,}000\ \Omega \times .000071\ \text{A}$

$E = 2.13\ \text{V}$

TO FIND VOLTAGE DROP FOR R_2

$E = R_2 \times I$

$E = 40{,}000\ \Omega \times .000071\ \text{A}$

$E = 2.84\ \text{V}$

Figure 1-9 Typical computer circuit. Note: Adding the two voltage drops does not precisely equal 5v since we were significantly to the sixth decimal place on the amperage calculation. Generally, it is not necessary to go any further and the .03v would be considered insignificant.

§ 1-5 PARALLEL CIRCUITS

Parallel circuits are generally defined as having more than one circuit path for the electrons to follow (Figure 1-10). Each extra path is called a branch circuit. When devices are wired in parallel, the operational characteristics differ from series circuits. There are three characteristics of parallel circuits you need to know:

- The voltage applied to, or measured across, each branch circuit is the same.
- The sum of the individual branch amperages equals the total amperage.
- The total resistance is less than the lowest branch resistance.

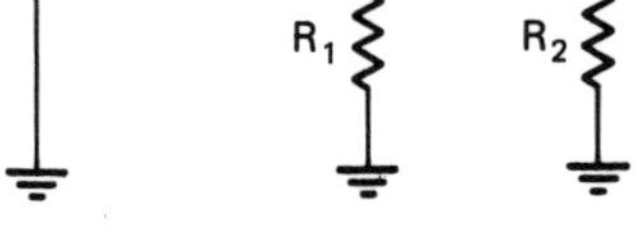

Figure 1-10 Parallel circuit.

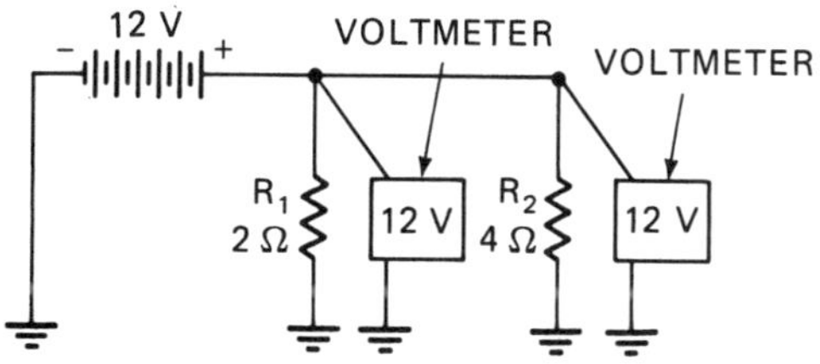

Figure 1-11 Voltage in parallel circuits. Since each branch is connected directly to battery + and battery –, each voltmeter will read 12v.

1-5-1 Voltage in a Parallel Circuit

The voltage applied to, or measured across, each branch circuit is the same (Figure 1-11). In this figure, if 12v is applied to each branch, 12v will be dropped across each branch.

1-5-2 Amperage in a Parallel Circuit

The sum of the individual branch amperages equals the total amperage (It). The branch amperages (each branch) must be calculated first, then added together for the total amperage.

To find each branch amperage: I = E/R.

To find total amperage: I1 + I2 = It. (If there are more branches they are added in). See Figure 1-12.

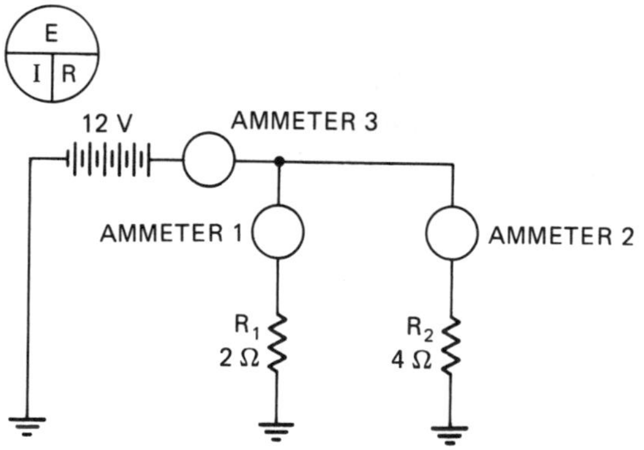

I (THROUGH R_1) = $\frac{12\text{ V}}{2\ \Omega}$ = 6 A (READING ON AMMETER 1)

I (THROUGH R_2) = $\frac{12\text{ V}}{4\ \Omega}$ = 3 A (READING ON AMMETER 2)

$I_t = 6\text{ A} + 3\text{ A}$

$I_t = 9\text{ A}$

Figure 1-12 Amperage in a parallel circuit. To find total amperage (I_t) $I_t = I_1 + I_2$

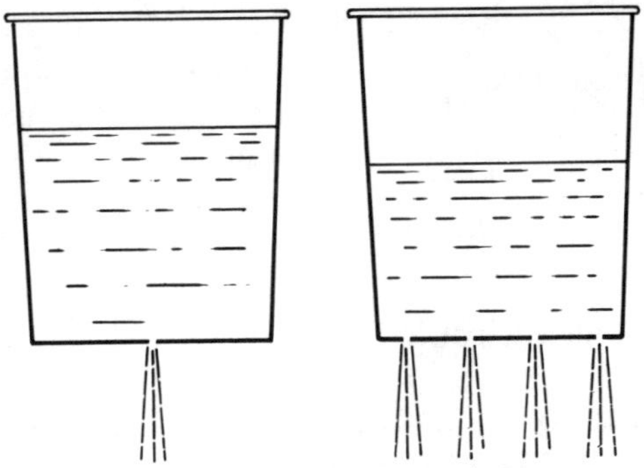

Figure 1-13 As the number of holes increases, the amount of water flowing out of the bucket increases.

1-5-3 Resistance in a Parallel Circuit

The total resistance is less than the lowest branch resistance. The reason for this is that as more branches are added, there are more paths for electrons to follow. As branches are added, current flow increases and resistance decreases.

To help you visualize, consider a bucket of water suspended in the air (Figure 1-13). If a small hole were drilled in it, only a small trickle of water would run out. If you add more holes, more water would run out—there are more holes now and, therefore, less resistance. The same is true of an electrical circuit. As more branches are added, current flow increases and resistance decreases.

There are different methods to find Rt. Perhaps the easiest method is the *assumed voltage* method, in which the sum of the branch amperages is divided into source voltage (Figure 1-14).

$$Rt = \frac{E}{It}$$

Another method for calculating total resistance (Rt) is as follows (see Figure 1-15):

$$Rt= \frac{1}{\frac{1}{R1} + \frac{1}{R2}}$$

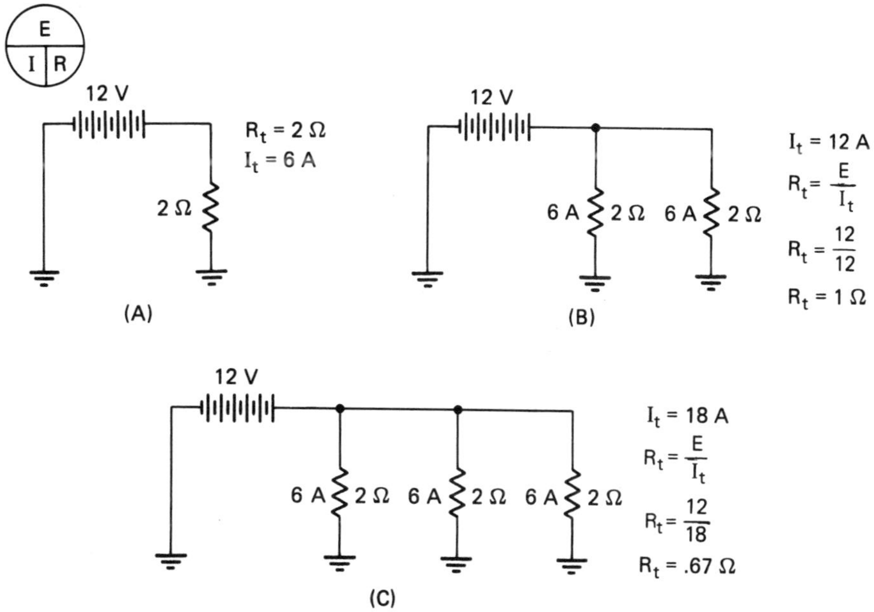

Figure 1-14 Assumed Voltage Method

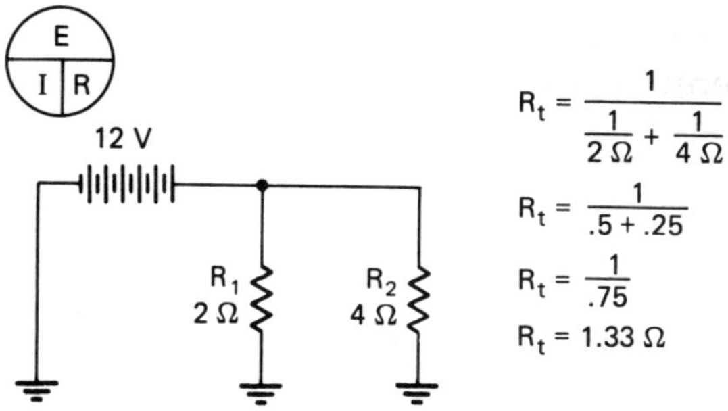

Figure 1-15 Another method to find total resistance (Rt).

If all the branches of a parallel circuit have the same resistance, then finding the total resistance is easy. The following formula will work (See Figure 1-16):

$$Rt = \frac{\text{Resistance of any branch}}{\text{Number of branches}}$$

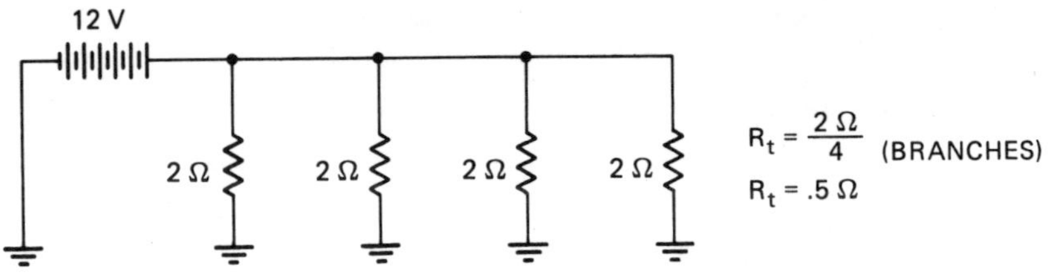

Figure 1-16 Finding total resistance when all branches have the same resistance.

§ 1-6 MAGNETIC FIELDS

Whenever current passes through a conductor, a magnetic field is created surrounding the conductor. The greater the current flow the greater the strength of the magnetic field. The magnetic field is made of lines of force. The more lines of force in a given area, the greater the magnetic field strength (Figure 1-17).

Some materials allow magnetic lines of force to pass easily through them. This property is called permeability. Iron is permeable to magnetic fields and is often used to concentrate the magnetic lines of force.

Materials that resist lines of force have a reluctance. Air for example has a high reluctance, (Figure 1-18).

Magnetic fields can induce a voltage in a conductor. Whenever the lines of force move through a conductor a voltage is induced.

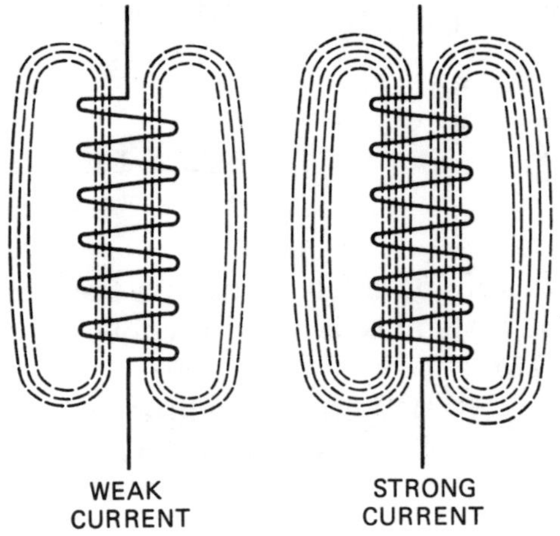

Figure 1-17 As current flow increases, magnetic field strength increases

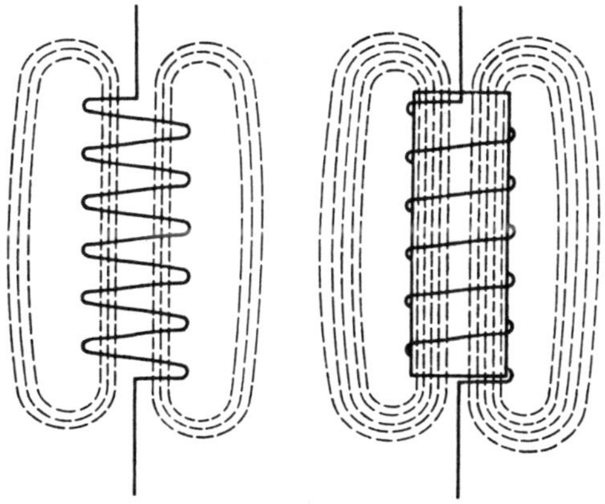

Figure 1-18 Iron is permeable to magnetic lines of force and will concentrate them. Air has a high reluctance to magnetic lines of force. Magnetic lines of force in the air tend to spread out more.

The strength of the induced voltage depends on the relative motion between the conductors and magnetic field strength. Relative motion can be the physical movement of the conductors or the expansion and contraction of the magnetic field.

Magnetic fields are used in a variety of applications, but this textbook is concerned with their use in a transformer. A transformer works by mutual induction where energy is transferred from one coil to another. In a simple two-coil setup, here is what happens: When there is no magnetic field around the primary coil, there is no energy transferred to the secondary coil. But when current first passes through the primary coil, the magnetic field expands. And as the magnetic lines of force expand and move through the secondary coil a voltage is induced. If the voltage remains constant in the primary coil, the magnetic field stabilizes and no current will be induced in the secondary coil. But should current flow stop in the primary coil, the magnetic field would collapse and voltage would again be induced in the secondary coil, Figure 1-19.

§ 1-7 CAPACITANCE

Capacitance, simply put, is the ability to hold a charge. A capacitor is a component containing opposing plates with a dielectric material between the plates built to hold a charge. One set of plates is connected to the positive side while the other set is connected to the

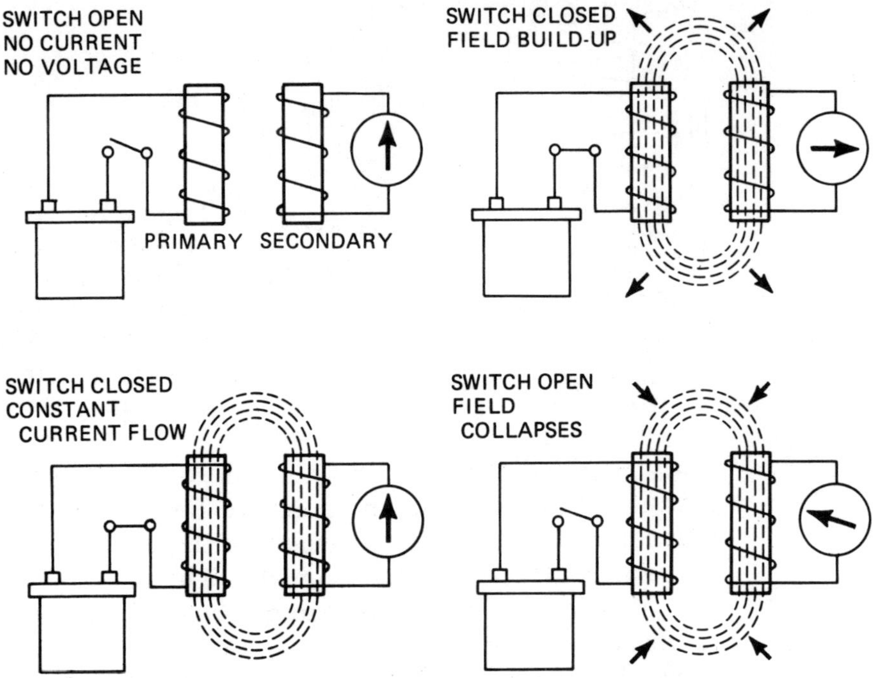

Figure 1-19 Transformer. Note the direction the voltmeter travels as the magnetic field expands and collapses. (Courtesy of Ford Motor Company.)

negative side. Separating the plates is a dielectric material (Figure 1-20). When charging a capacitor the positive plates contain very few free electrons while the negative plates have an abundance of

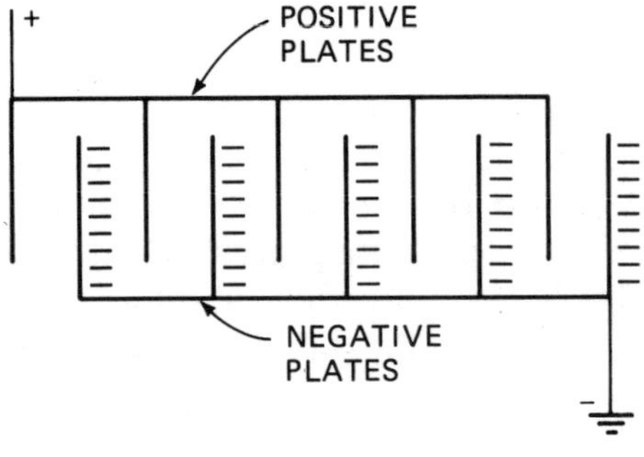

Figure 1-20 Capacitor.

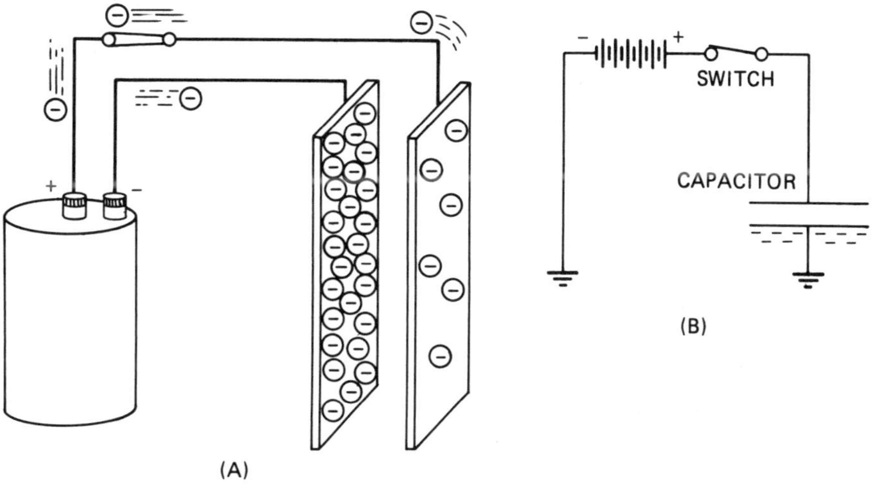

Figure 1-21 (a) Electrons are attracted to the + side of battery. Electrons are repelled by the – side of battery. (Courtesy of Ford Motor Company.) (b) Capacitor charging. When the switch is closed, current will flow until the capacitor is charged; that is, it will hold all the electrons its size will permit.

free electrons (Figure 1-21). If the circuit were opened the capacitor would hold its charge until the circuit becomes closed again.

There are three factors governing capacitance:

- distance between opposing plates
- type of dielectric material
- amount of plate surface area

To increase capacitance you need to decrease the distance between opposing plates, increase the amount of plate surface area, and use a dielectric material that will not leak easily.

§ 1-8 ANALOG AND DIGITAL SIGNALS

A common characteristic of analog signals is that they are continuously variable. For example, voltage across a variable resistor in a series circuit would change as the resistance changes. Any change in resistance would immediately result in a change in voltage.

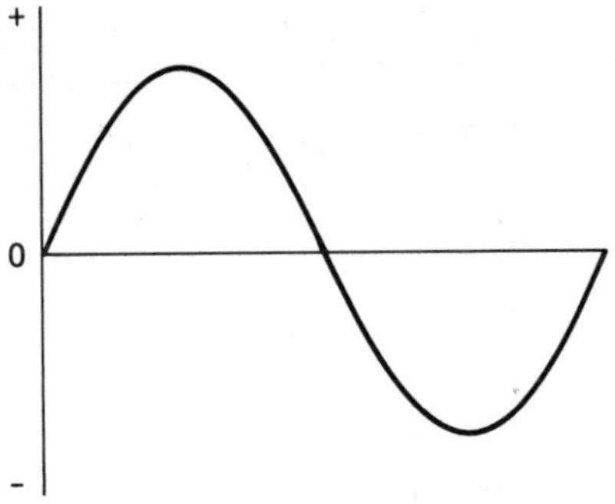

Figure 1-22 One AC cycle.

Analog signals also have a frequency and amplitude. The frequency of an analog signal is determined by the number of cycles per second. A cycle is a process that begins where it ends. In electricity, a cycle can be represented with a sine wave. In Figure 1-22, a voltage signal begins to move upward when it has a positive polarity, then downward with a negative polarity. As the voltage signal ascends again, it reaches its beginning point.

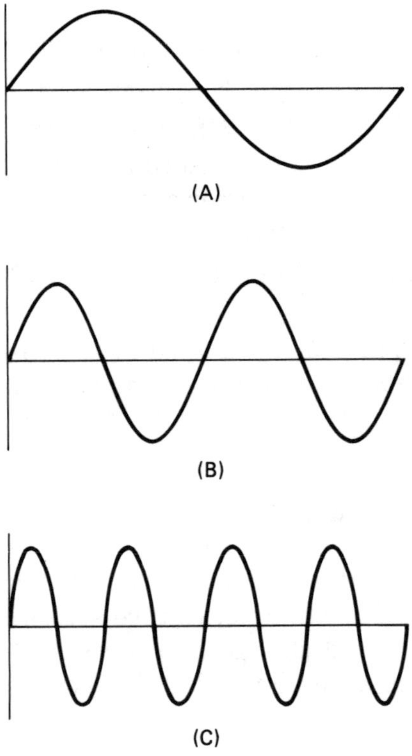

Figure 1-23 Frequencies. (a) 1 second, 1 cycle = 1Hz./second, 2 cycles = 2Hz./second, 4 cycles = 4 Hz.

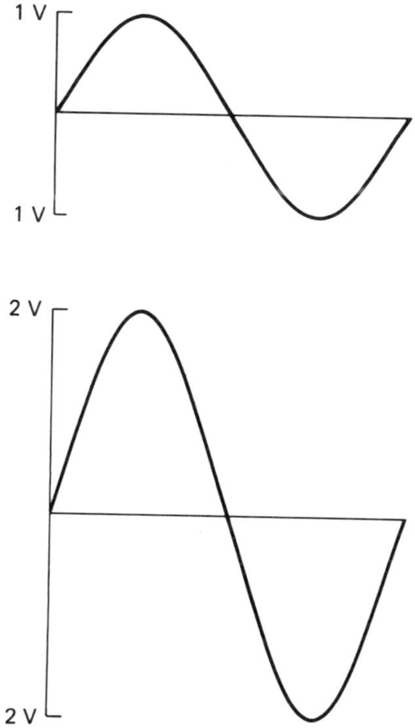

Figure 1-24 Note that the amplitude changed, not the frequency.

One cycle per second equals one hertz (Hz). The number of cycle repetitions per second is the frequency (Figure 1-23). For example, one thousand cycles per second (1,000 Hz) equals one kilohertz (1kHz) or one kilocycle per second.

The amplitude is the peak value of the sine wave. This value represents the sine wave's intensity. Two voltage signals can have the same frequency but can also have different amplitudes (Figure 1-24).

Digital signals are variable but not continuously, and they come in discrete units. A digital signal is represented by a voltage/no voltage condition or a high voltage/low voltage condition. This forms a square wave (Figure 1-25). The digital signal is useful in indicating on/off, yes/no, and high/low conditions. A digital system may be referred to as a binary system.

§ 1-9 DIODES

Diodes are semiconductor, solid state components that allow current flow in one direction and block current flow when the direction

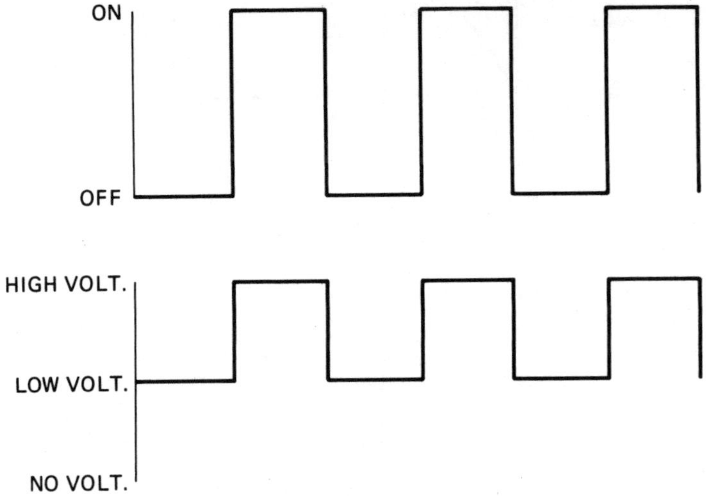

Figure 1-25 Digital signals.

is reversed. The arrow in the diode symbol indicates the direction in which the diode will allow current to flow (Figure 1-26). Diodes used in alternators change alternating current to direct current. In computer circuits, diodes are often used to prevent voltage spikes.

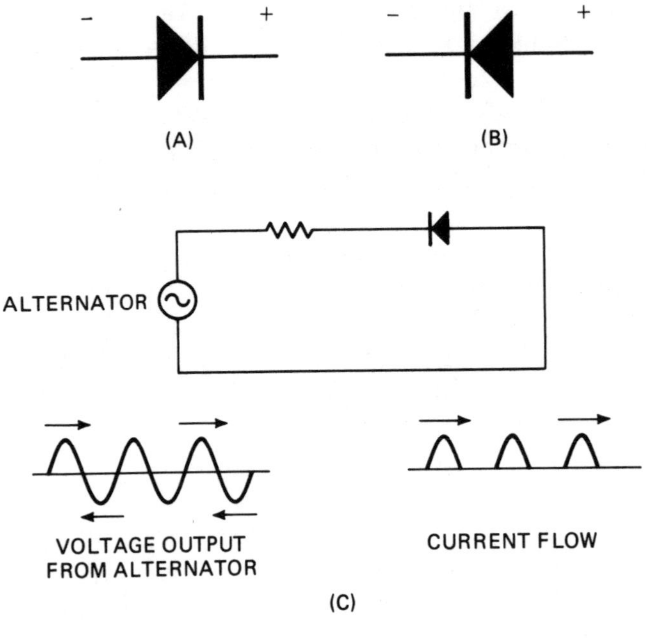

Figure 1-26 Diode. (a) No current flow. (b) Current flow. For a diode to conduct in a circuit, the arrow in the symbol *must* point toward the negative side of the circuit. (c) Diode used in alternator.

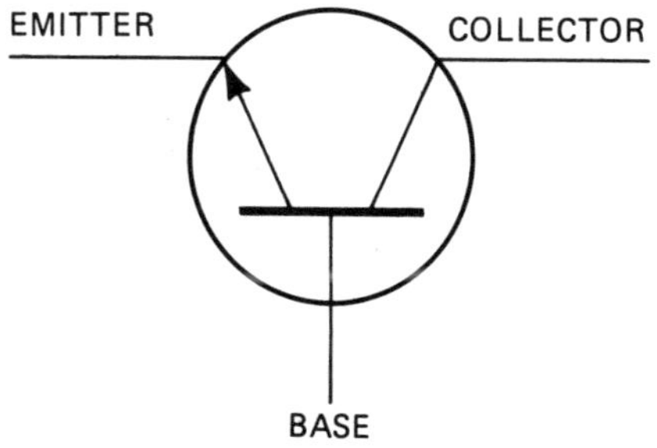

Figure 1-27 A transistor.

§ 1-10 TRANSISTORS

Transistors are semiconductor, solid state components which are often used as switching devices, relays, and amplifiers. The transistor is made of three components: base, collector, and emitter (Figure 1-27). Essentially, when a very tiny current is applied to the base the base becomes conductive allowing a much larger current to flow in the collector/emitter circuit (Figure 1-28). When current

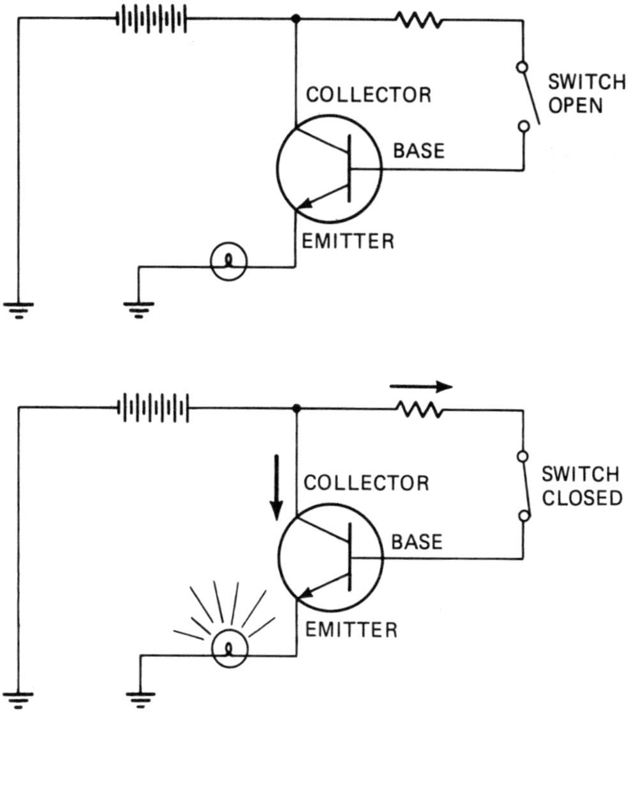

Figure 1-28 In the top view, the switch is open and no current is flowing. In the bottom view, the switch is closed, making the base of the transistor conductive. The majority of current will flow from the collector/base/emitter to the light-bulb path.

is removed from the base the base is no longer conductive and no current will flow through the emitter and collector.

QUESTIONS

1-1. Define amperage, voltage, and resistance.

1-2. Convert the following . . . to numbers. 30ma, 450mv, 2M, 14kv

1-3. Convert the following to the appropriate prefix. .750v, 10,100,000, .045a, 7000v

1-4. Write Ohm's law.

1-5. Describe what happens to current flow if the voltage increases and resistance remains the same.

1-6. Describe what happens to current flow if resistance increases and the voltage remains fixed.

1-7. Write the three mathematical expressions of Ohm's law.

1-8. Write the three characteristics of a series circuit.

1-9. Write the three characteristics of a parallel circuit.

1-10. Describe transformer operation.

1-11. Define capacitance and three factors governing capacitance.

1-12. Describe analog signal.

1-13. Define amplitude and frequency.

1-14. Describe digital signal.

1-15. Describe diode operation.

1-16. Describe transistor operation.

Chapter 2

The Computer and Circuit Operation

OBJECTIVES

After studying this chapter, the student will be able to

- briefly describe computer system operation
- state the different types of input and output circuits
- describe the operation of thermistors, potentiometers, manifold absolute pressure sensors, magnetic reluctance sensors, Hall effect sensors, exhaust gas oxygen sensor, knock sensor, permanent magnet sensor, and digital switch inputs
- describe why corrosion affects computer circuits
- state the components of the computer and describe their functions
- describe the operation of solenoids, relays, motor outputs
- describe the difference between normally closed and open solenoids
- describe ground side switching
- describe the purpose of the clamping diode

§ 2-1 COMPUTER SYSTEM OVERVIEW

Fuel control, ignition timing, and emission control are just some of the functions monitored and controlled by the computer. In order to understand how the computer manages these functions, computer circuit operations must be understood.

Automotive computer systems can be divided into three main sections:

- input (sensors)
- computer
- output (actuators)

The input section is the information gathering portion of the computer circuit. Inputs exist to provide information to the computer about engine operating conditions and the environment. Inputs convert physical properties and mechanical motion into electrical signals the computer will understand. The computer processes input information and arrives at a decision based on its programming. The outputs, controlled by the computer's decisions, modify engine operation. Figure 2-1 shows a simplified example. The engine coolant temperature sensor signals the computer that

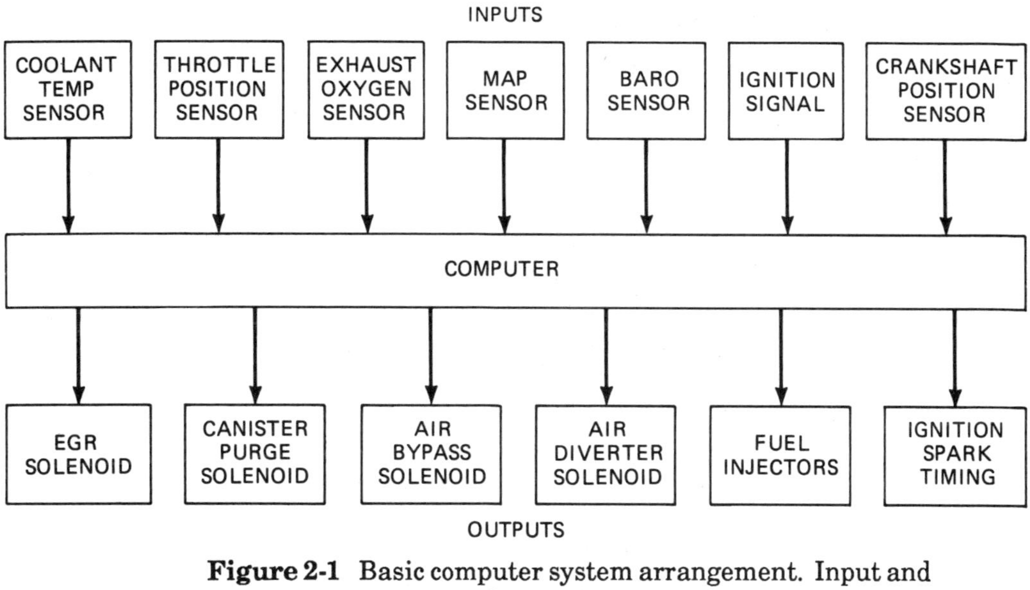

Figure 2-1 Basic computer system arrangement. Input and output usage varies with specific systems.

the engine is cold. The computer takes and processes the information. The computer now realizes the engine is cold and according to its programming it will signal the fuel injectors to stay open longer allowing extra fuel to enter the engine to provide for easier starting. The inputs, the computer, and the outputs all work together to control engine operation. How they work is the subject of the next three sections.

§ 2-2 INPUTS (SENSORS)

The computer needs information on the following:

- engine coolant temperature
- air temperature
- throttle position
- exhaust gas oxygen content
- engine speed
- engine crankshaft position
- intake manifold pressure
- barometric pressure
- engine detonation
- transmission gear position
- brakes on or off
- air conditioning on or off
- vehicle speed
- EGR valve position
- amount of air entering the engine

Please note that not all manufacturers use all the inputs shown above on their computer systems. To send this information to the computer various types of sensors are used. Each is discussed below.

2-2-1 Thermistor Type (Variable Resistance) Sensors

A thermistor is a variable resistor sensitive to temperature. As the temperature of the thermistor changes, its resistance value chan-

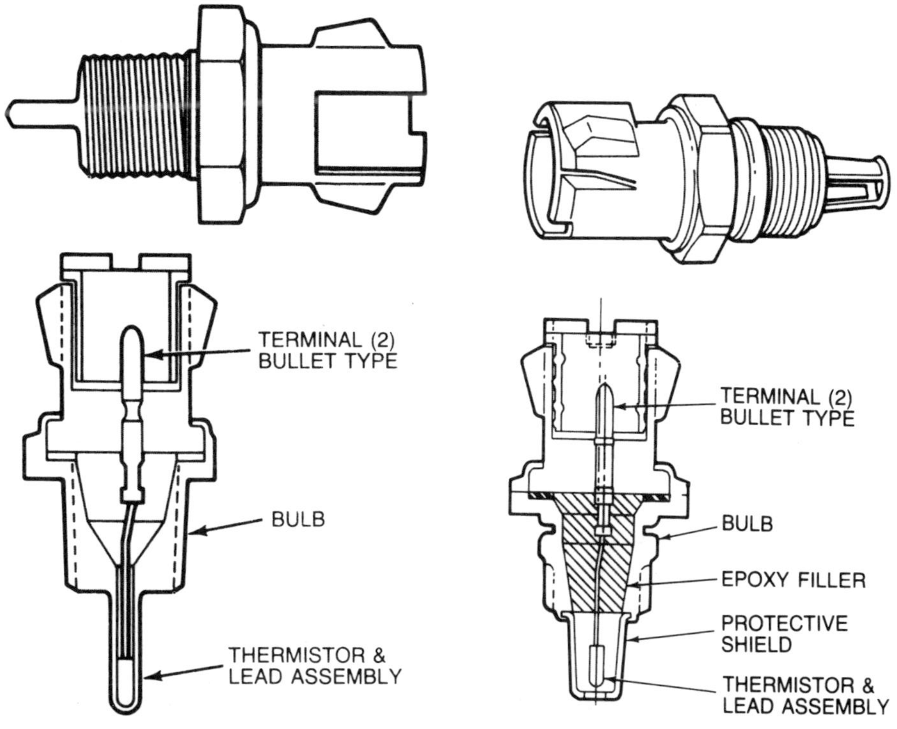

Figure 2-2 (a) Coolant temperature sensor. (b) Air temperature sensor. (Courtesy of Ford Motor Company)

ges. Thermistors are used to sense engine coolant, intake air, and air temperatures (see Figure 2-2). Thermistors are made of a negative temperature coefficient material (NTC). When made of an NTC material, the thermistor resistance value decreases as temperature increases.

The change in resistance value affects the electrical signal sent by the computer. In a typical circuit such as the engine coolant sensor (see Figure 2-3), an electrical signal of five volts passes through a current limiting resistor (which has a fixed resistance) located inside the computer, to the sensor (variable resistance), and then back to the computer to ground. Connected between the current limiting resistor and the coolant temperature sensor is an analog to digital signal processor (also called an analog to digital converter). The A/D signal processor will measure the voltage drop across the coolant sensor.

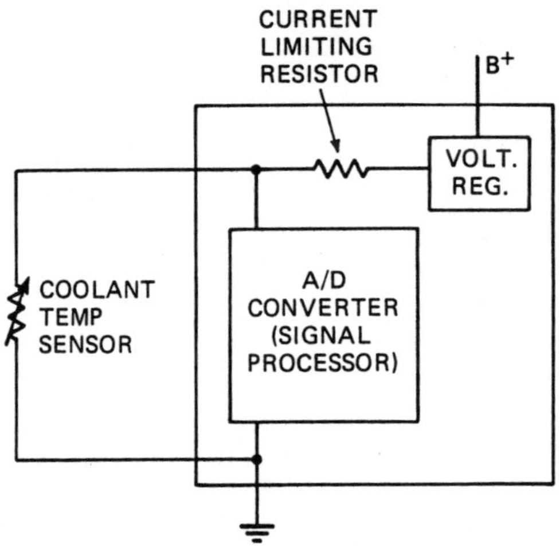

Figure 2-3 Coolant temperature sensor circuit.

When cold, the resistance value for the sensor is high and therefore the voltage drop is high. This high voltage drop indicates to the computer that the engine is cold. As the engine warms up, sensor resistance decreases and the voltage drop is less, indicating to the computer that the engine is getting warmer. See Figure 2-4.

The other thermistor sensors operate in the same way, though their operating range and reaction to temperature changes are different.

2-2-2 Potentiometer (Variable Resistance) Sensors

A potentiometer is a variable resistor that is used by the computer to indicate position and direction of movement of a device. All potentiometers have three wire connections. One wire provides a steady voltage to one end of the resistor and the other end is connected to ground. The third wire is the signal wire which makes electrical contact with the resistor but it is free to glide over the resistor. As the signal wire moves over the resistor the voltage in the signal line changes, Figure 2-5.

Throttle position sensors (TPS) are usually potentiometers. The computer needs to know at all times the position of the throttle and the direction of movement. The throttle shaft is mechanically connected to the signal wire. Of the other leads, one provides a constant voltage of five volts from the computer while the other lead returns to the computer to ground. As the throttle shaft

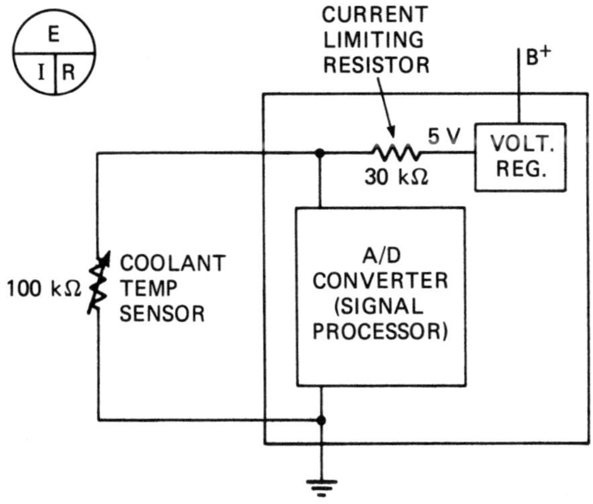

R_t = 100 kΩ + 30 kΩ = 130 kΩ

$I = \frac{5}{130\ k}$ = .000038 = .038 ma

E (COOLANT SENSOR) = .038 ma × 100 kΩ = 3.8 V

(A)

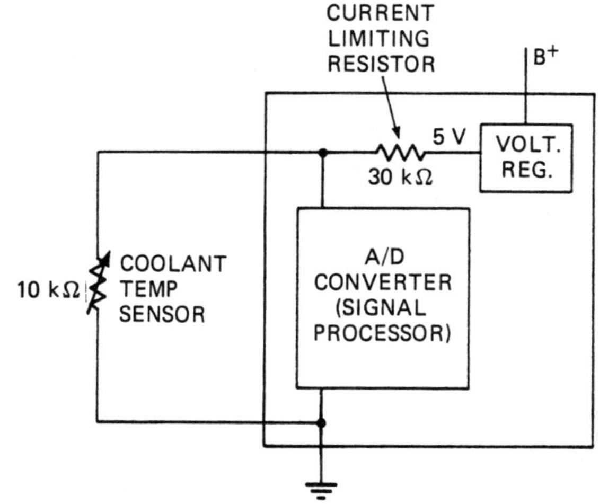

R_t = 10 kΩ + 30 kΩ = 40 kΩ

$I = \frac{5}{40\ k\Omega}$ = .000125 = .125 ma

E (COOLANT SENSOR) = .125 ma × 10 kΩ = 1.25 V

(B)

Figure 2-4 Thermstor sensor. (a) engine cold. (b) engine warm.

rotates, the voltage in the signal line changes, indicating to the computer the amount the throttle valve has opened and whether it is opening or closing. See Figure 2-6.

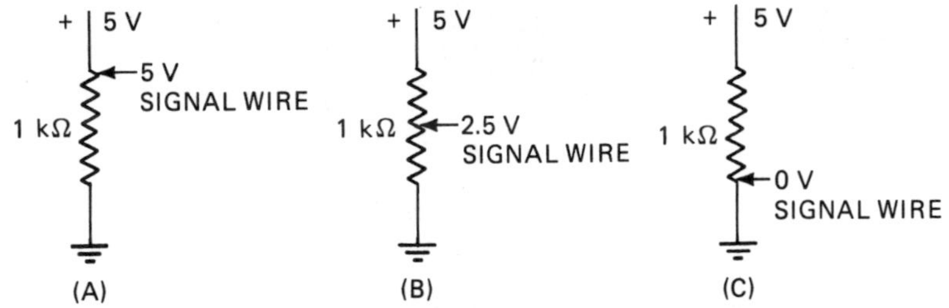

Figure 2-5 Total voltage drop across the resistor is 5v. The signal wire reads the voltage drop remaining.

Another potentiometer is the EGR valve position sensor. The purpose of this sensor is to indicate to the computer the amount the EGR valve is opening by electrically measuring valve height. The EGR valve stem is connected to the signal line and its operation is exactly the same as the throttle position sensor since they are both potentiometers, Figure 2-7.

A third potentiometer is the vane air flow meter with the vane mechanically connected to the signal line. Vane position is determined by the velocity of the intake air. As the throttle valve opens and engine RPM's increase, intake air velocity increases moving the vane and therefore changing the voltage signal, Figure 2-8.

2-2-3 Exhaust Gas Oxygen (EGO) Sensor

The computer needs information on how lean or rich an air/fuel mixture is being fed to the engine. An exhaust gas oxygen sensor measures the exhaust gas oxygen content. From the oxygen content a voltage signal (anywhere from 0 to 1 volt) is generated and sent to the engine computer (see Figure 2-9). The EGO sensor is classified as a signal generator.

The EGO is principally made up of zirconium dioxide (ZrO_2) which will attract negatively charged oxygen ions. One part ZrO_2 is exposed to atmospheric oxygen while the other half is exposed to exhaust gas oxygen. On each side of the ZrO_2 element is a platinum electrode that completes the circuit through the computer, Figure 2-10. When the exhaust gases are low in oxygen content because of a rich air/fuel ratio, the voltage signal is high. The reason is that there are far more oxygen ions and therefore more electrons on the atmospheric side of the EGO sensor than on the exhaust side. This

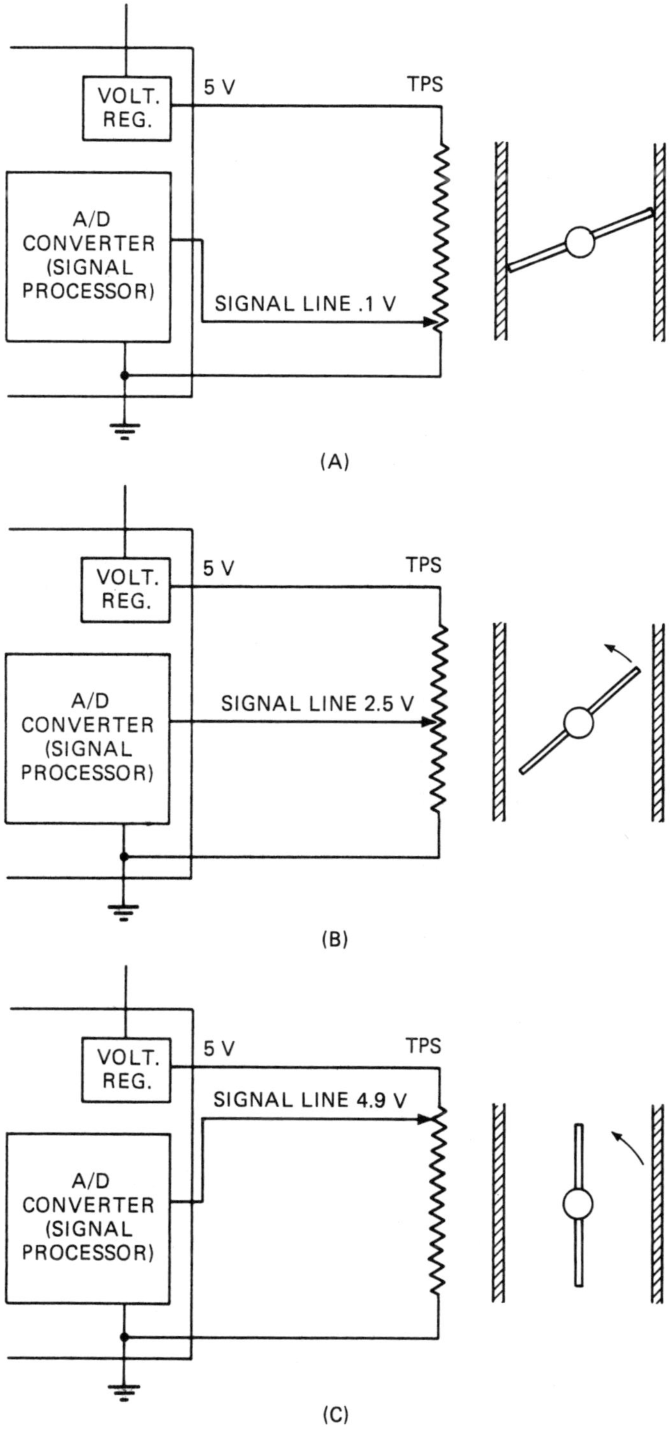

Figure 2-6 Throttle position sensor. (a) Throttle at idle. (b) Throttle at half. (c) Throttle wide open.

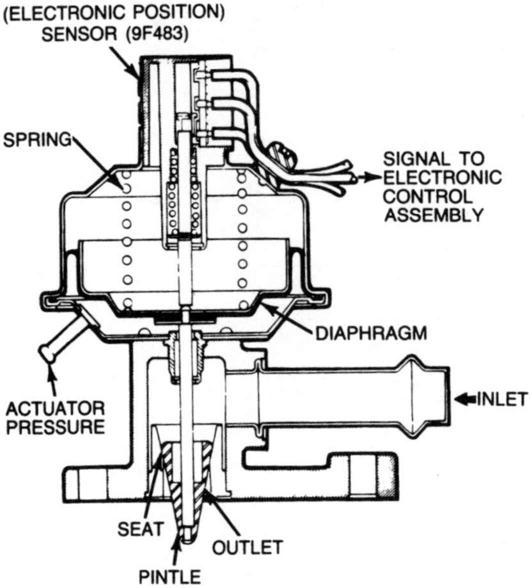

Figure 2-7 EGRVP sensor. (Courtesy of Ford Motor Company)

imbalance creates a high voltage signal. The greater the imbalance in oxygen content the higher the voltage signal. This range is from 600 to 900 millivolts (see Figure 2-11). When the exhaust gasses are high in oxygen content because of a lean air/fuel ratio, the voltage signal is low. This time there are more oxygen ions and therefore more electrons on the exhaust side. The electrical imbalance between atmosphere and exhaust is small, and the resultant voltage signal will be low. This range is from 100 to 400 millivolts (see Figure 2-12). When the air/fuel ratio is stoichiometric (that is the chemically correct ratio of air to fuel—14.7 to 1), the voltage signal is approximately 450 millivolts (see Figure 2-13).

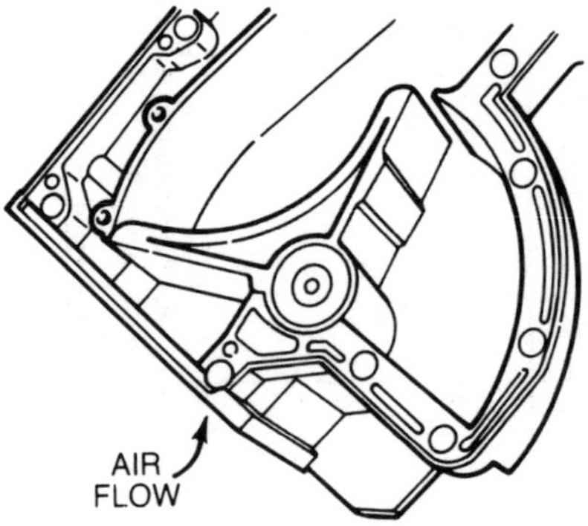

Figure 2-8 Vane air flow meter. (Courtesy of Ford Motor Company)

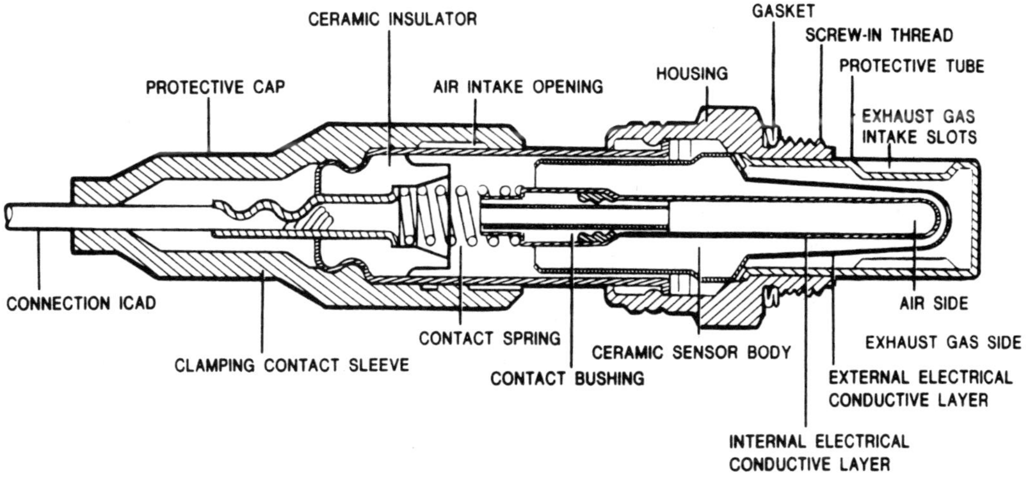

Figure 2-9 Exhaust gas oxygen sensor. (Courtesy of Ford Motor Company)

One characteristic of ZrO_2 is that it must be above 300°C (572°F) in order to generate an accurate voltage signal. When the oxygen sensor is cold the engine computer ignores the EGO and it must wait until the EGO warms up. To speed up the warm-up period heated EGO's (HEGO) have been developed (see Figure 2-14). An

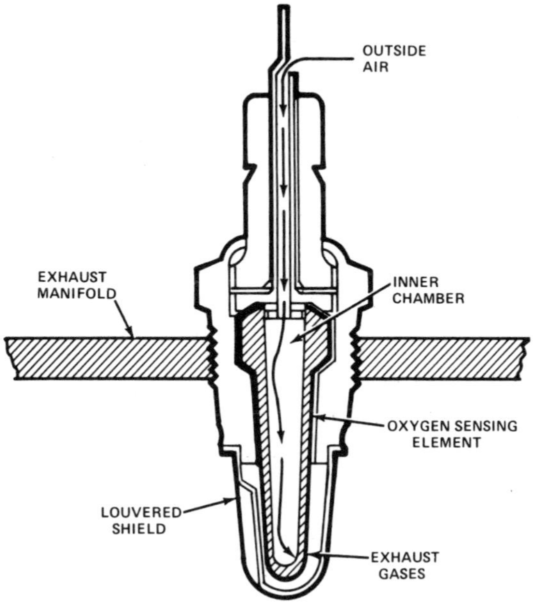

Figure 2-10 (Courtesy of General Motors Product Service Training)

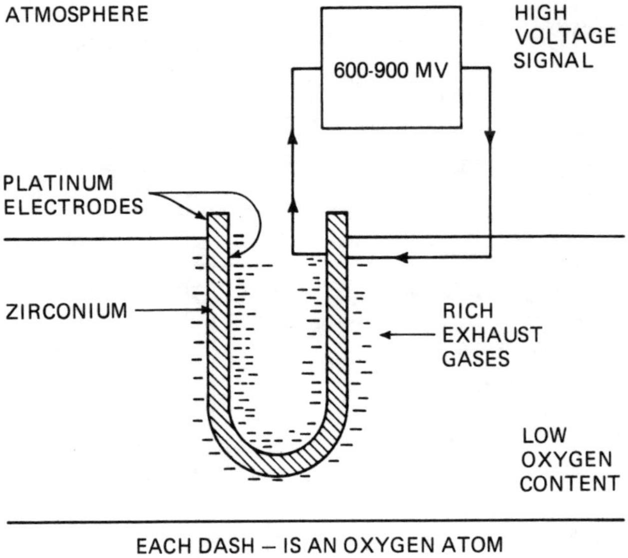

Figure 2-11 Exhaust gas oxygen sensor (rich air/fuel ratio.

electrical resistance element is installed inside the HEGO and operates off battery voltage. The element is a positive temperature coefficient type (PTC) which increases in resistance as temperature increases, reducing amperage through the element. Voltage signals can now be produced when the exhaust gas temperature is only 200°C (392° F).

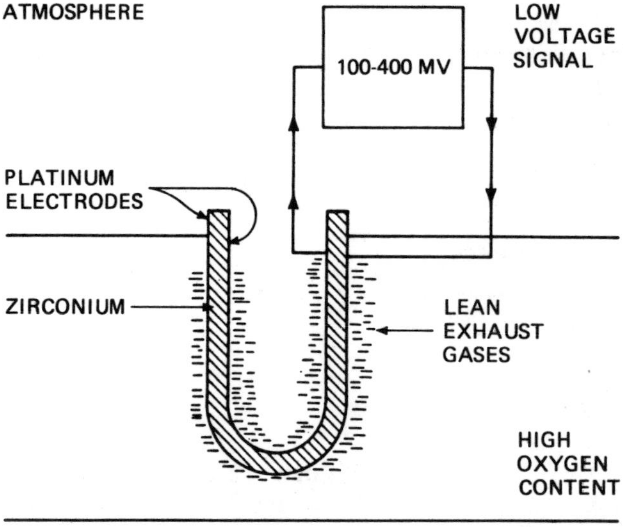

Figure 2-12 Exhaust gas oxygen sensor (lean air/fuel ratio).

RICH MIXTURE → LOW OXYGEN CONTENT → HIGH VOLTAGE SIGNAL

14.7:1 → STOICHIOMETRY → .450 MV

LEAN MIXTURE → HIGH OXYGEN CONTENT → LOW VOLTAGE SIGNAL

Figure 2-13 Correlation between air/fuel mixture, exhaust gas oxygen content, and oxygen sensor voltage output.

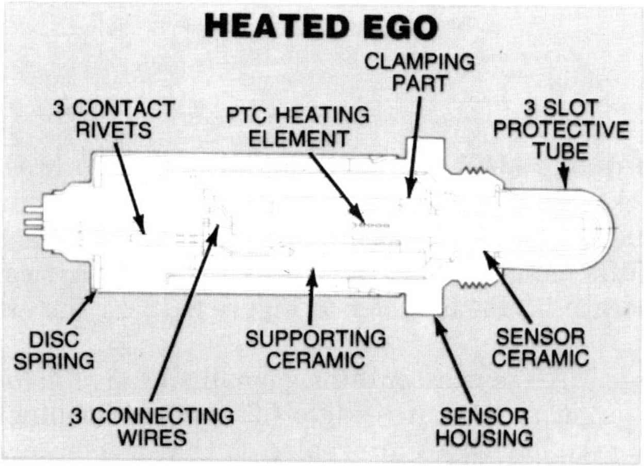

Figure 2-14 (Courtesy of Ford Motor Company)

Based on the input from the EGO sensor, the computer can make the necessary adjustments to the fuel system to provide the most ideal air/fuel ratio possible.

2-2-4 Manifold Absolute Pressure (MAP) and Barometric Pressure (BP) Sensors

The MAP sensor is used to measure intake manifold pressure (vacuum). This sensor provides essential information needed by the computer to control fuel and spark advance (see Figure 2-15).

All MAP sensors are connected to intake manifold pressure by a plastic or rubber hose. There are three major types of MAP sensors:

- strain gauge (Wheatstone bridge)
- capacitive capsule
- linear variable differential transformer (LVDT)

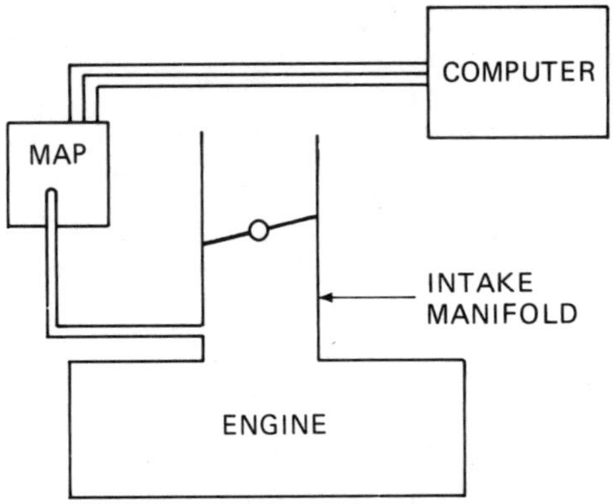

Figure 2-15 MAP sensor.

2-2-4-1 Strain gauge MAP sensor. The strain gauge MAP sensor is based on the Wheatstone bridge principle. Wheatstone bridge circuits are used in many applications to produce a voltage signal proportional to change in electrical resistance, or physical movement, or temperature. Please refer to Figure 2-16 for the explanation.

The strain gauge MAP sensor contains a small silicon chip (or diaphragm) that is larger at the outer edges (.25mm) and thinner at the center (.025mm). The edges are sealed so that a reference chamber is formed on one side of the silicon chip creating a diaphragm (see Figure 2-17). At each edge the silicon chip is doped with an impurity to form a piezoresistor. A piezoresistor changes its resistance with changes in pressure. The piezoresistors are connected to form a Wheatstone bridge. On the opposite side of the reference chamber there is an opening for intake manifold pressure. When there is no strain on the diaphragm all four piezoresistors have an equal value, and therefore no voltage signal is present. However, when the silicon diaphragm is strained, the resistance value of the piezoresistors changes and an imbalance is created across the Wheatstone bridge, generating a voltage signal. This voltage signal is sent to an amplifier which in turn will send a signal to the computer in proportion to intake manifold pressure (see Figure 2-18).

2-2-4-2 Capacitor-Capsule MAP Sensor. This MAP sensor is essentially a variable capacitor. The capacitor capsule is constructed of two silicon plates separated by a sealed chamber (see Figure 2-19). On each plate an electrode is connected to two silicon plates which are designed to flex inward under manifold pressure.

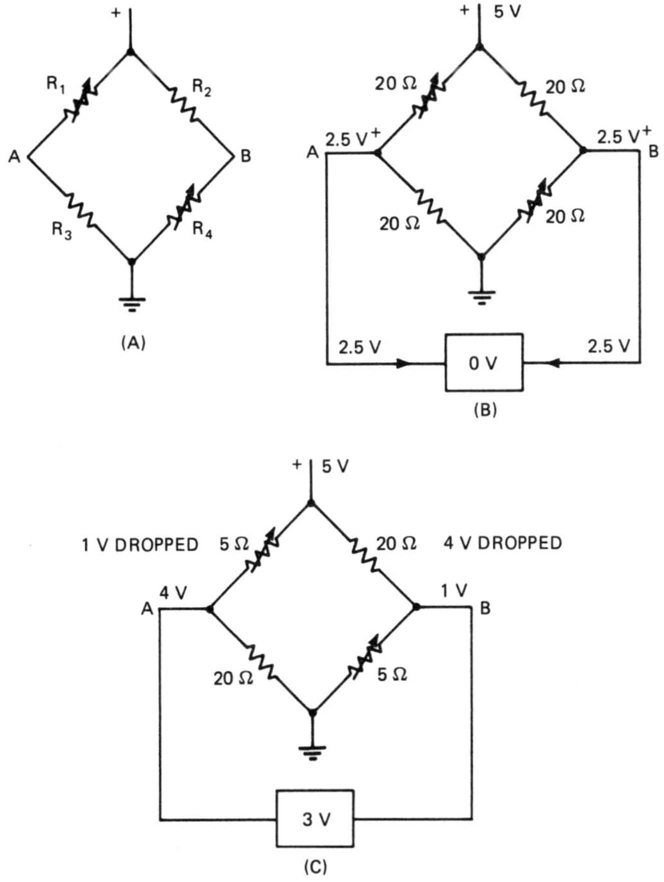

Figure 2-16 The Wheatstone bridge (a) consists of two parallel branches with each branch having two resistors in series (left circuit is R1-R3, right circuit is R2-R4). R2 and R3 are fixed resistances. R1 and R4 are variable resistances attached to a mechanical device that can cause them to change their resistances. (b) As stated in Chapter 1, resistances in series divide up applied voltage. If the resistances are equal, the two resistors in each branch will divide the voltage equally. Therefore the voltage at points A and B will be the same; in this example, 2.5v. With the bridge balanced, the equal voltages produce an output of 0 volts. If a voltmeter were connected at points A and B, the voltmeter would read 0. (c) As a mechanical force causes the resistance values to change as shown, the voltage at A is 4 volts, voltage at B is 1 volt. The difference from A to B is the output signal of 3 volts.

As they flex inward toward each other capacitance increases as the distance decreases (review Chapter 1 on capacitance). The change in capacitance is detected and sent as a voltage signal to the computer.

2-2-4-3 Linear Variable Differential Transformer (LVDT) MAP Sensor. The LVDT MAP sensor is designed to give an output signal in proportion to physical displacement. The LVDT is constructed of one primary winding, two secondary windings wound in opposite directions, and a movable magnetic core. The magnetic

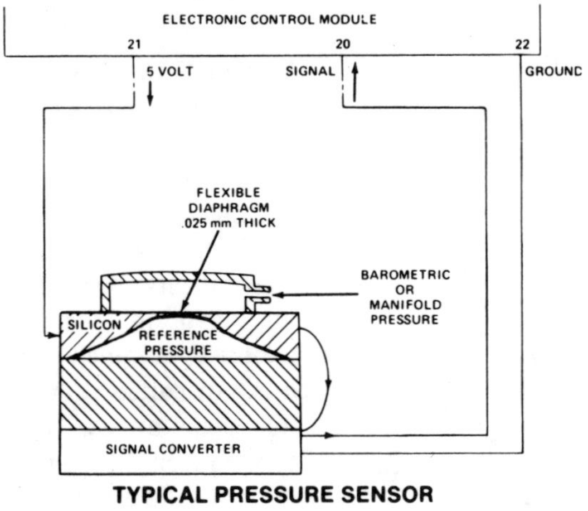

Figure 2-17 MAP sensor. (Courtesy of General Motors Product Service Training)

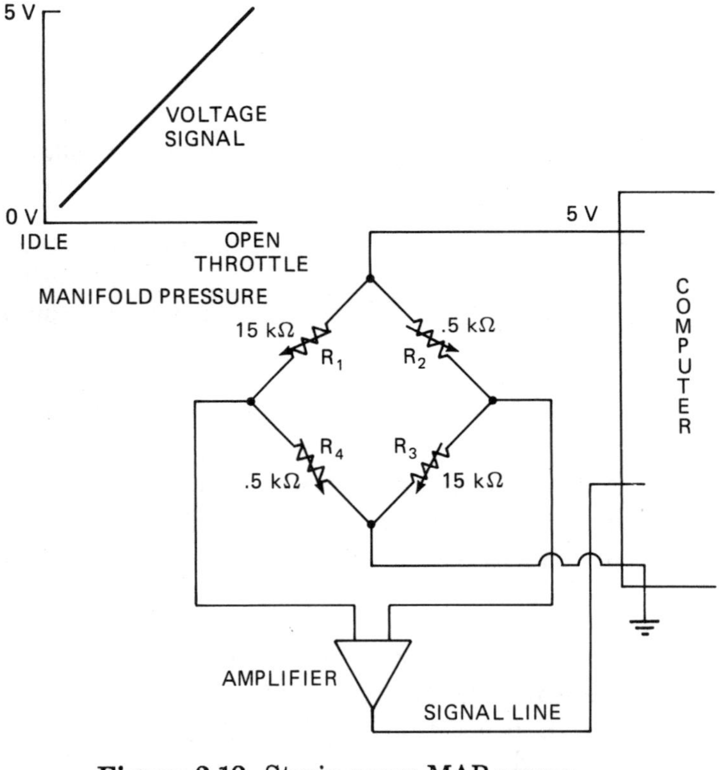

Figure 2-18 Strain gauge MAP sensor.

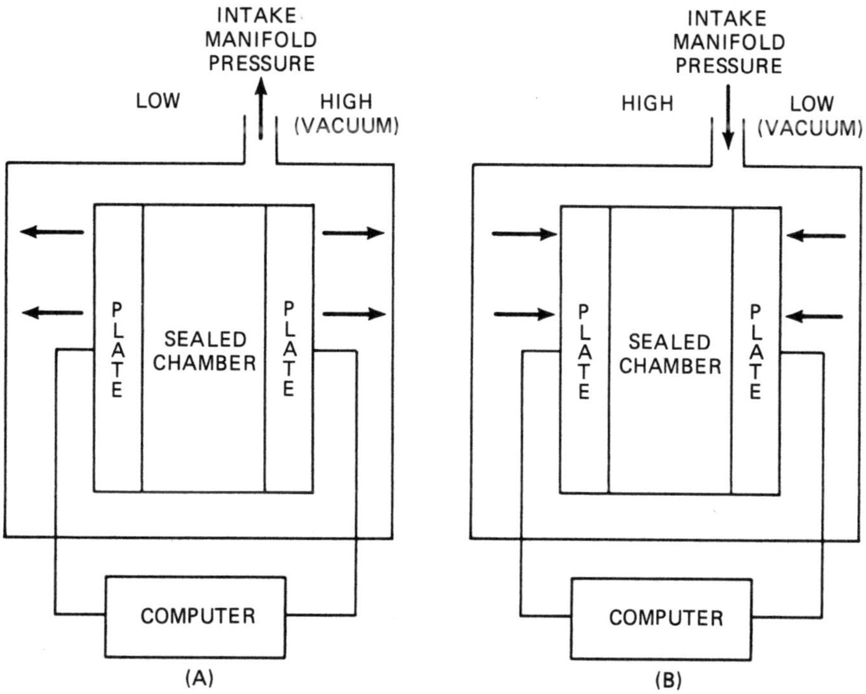

Figure 2-19 Capacitive capsule.

core is attached to an aneroid chamber which compresses as intake manifold pressure increases (see Figure 2-20). The primary winding at all times has a constant ac voltage applied to it. When the core is centered the voltage output between the secondary windings is zero. When the magnetic core moves, more magnetic lines will cut through one of the two secondary windings, creating a greater voltage in one of the windings. The imbalance in voltage between the two secondary windings increases in direct proportion to magnetic core movement; and movement of the core is in direct proportion to intake manifold pressure (see Figure 2-21). The voltage signal generated by the secondary windings is modified and sent to the computer.

Barometric sensors operate the same as MAP sensors but instead of measuring intake manifold pressure they measure atmospheric pressure.

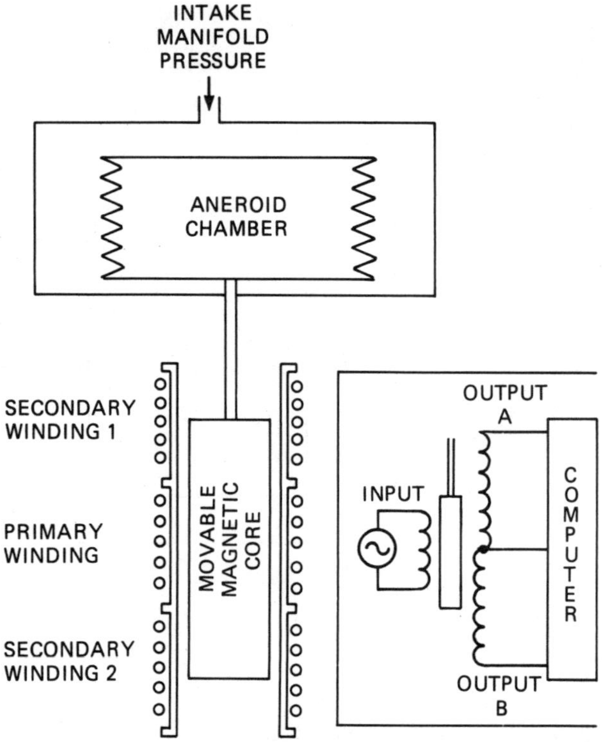

Figure 2-20 Linear variable differential transformer MAP sensor. Voltage output of the secondary windings is equal. Since they are wound in opposite directions, they will be of opposite polarity, and the difference in voltage between points A and B will be zero.

2-2-5 Variable (Magnetic) Reluctance Sensors

Variable reluctance (sometimes called magnetic reluctance) sensors come in a variety of forms and applications, but the basic operation is the same. The major components of this sensor are a sensing coil, permanent magnet, and a toothed or notched iron or steel sensing disc. The basic construction is shown in Figure 2-22.

When the teeth of the sensing disc are away from the magnetic pole piece, the magnetic field in the sensing coil is weak. The reason the magnetic field is weak is that the air gap has a high reluctance. As a tooth moves nearer to the pole piece the air gap decreases and the magnetic field concentrates, moving inward. As the magnetic field moves inward its lines of force cut through the sensing coil inducing a voltage. When the tooth is directly opposite the pole piece, the magnetic field reaches its highest concentration and the induced voltage is zero. As the tooth moves away from the pole piece, air gap reluctance increases and the magnetic field expands, inducing a voltage in the opposite direction in the sensing coil. Eventually the

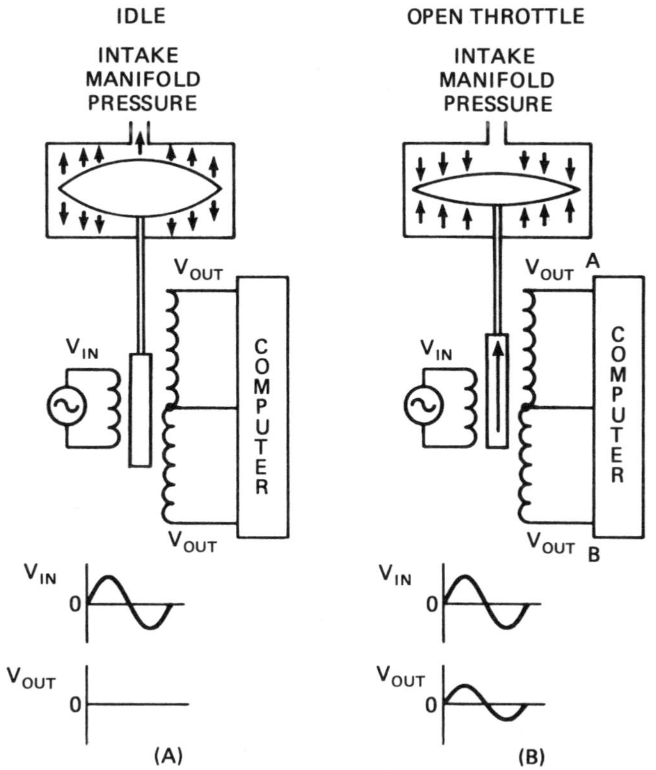

Figure 2-21 (a) Equal voltage signal sent to the computer results in zero volts. (b) Difference in voltage signals sent to the computer results in a voltage reading. Voltage at A is higher than that at B.

air gap is so great that the magnetic field becomes weak and stabilizes and therefore no more voltage is induced in the sensing coil (see Figure 2-23). The induced AC voltage signal can be used to trigger a variety of devices. It is commonly found in the distributor of electronic ignitions and it is used to tell the ignition module when to turn off primary current flow through the coil.

In engine computer circuits, the variable reluctance sensor is used to determine crankshaft position and engine rpm, and it is usually called a crankshaft position sensor (CPS) (see Figure 2-24). The sensing disc is generally mounted on the harmonic balancer, and it has just one-half the number of teeth as there are engine cylinders. The sensor probe position is fixed just above the sensing

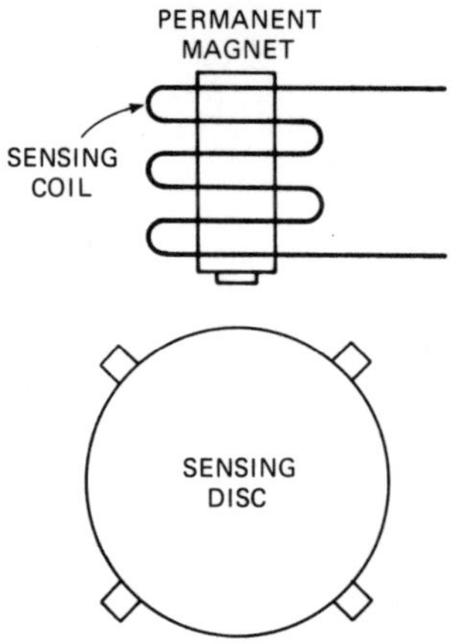

Figure 2-22 Magnetic reluctance sensor.

disc. The computer counts the number of induced voltage pulses as the teeth pass by the probe over a predetermined period. From this number the computer is able to determine crankshaft position and engine rpm. The computer will use the derived information to determine spark advance.

2-2-6 Hall Effect Sensors

Hall effect sensors are used to signal an ignition module, indicate engine rpm and crankshaft position. Hall effect sensors use a special rectangular semiconductor called a Hall element. The Hall element has a very unique property under certain conditions. When a current is passing through the element and magnetic lines of force are passing through the element at a right angle to current flow, a voltage is generated at the edges of the element which is perpendicular to current flow and the magnetic lines of force (see Figure 2-25). This is the Hall effect. If current flow is kept constant then any change in the lines of force will change the voltage output.

Typically, the Hall effect sensor is located in the distributor. A permanent magnet with a small air gap is placed opposite the Hall effect element. Fastened to the distributor shaft are rectangular metal shutters, or vanes, which fit in the air gap without touching the magnet or Hall element. When the shutter is not present in the air gap, the magnetic lines of force pass through the Hall element creating a small voltage output (see Figure 2-26a). When a shutter is in the air gap, the magnetic lines of force no longer reach the Hall element and the voltage is nearly zero, Figure 2-26b. The change in voltage output controls a switching transistor.

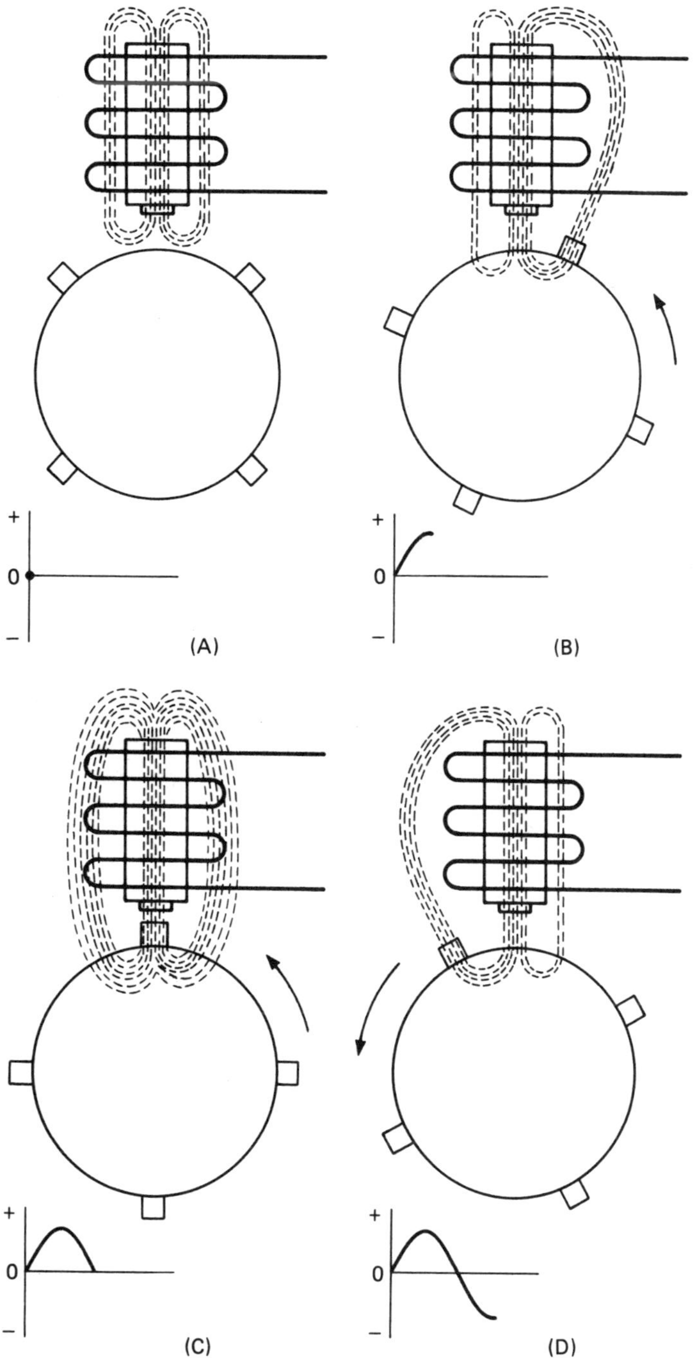

Figure 2-23 (a) Stationary magnetic field. (b) As the tooth approaches the magnet the air gap decreases and field strength increases. This increase or movement of the field caused a voltage to be produced in the coil. Voltage output and direction are indicated on the graph. (c) When the tooth lines up with the magnet, the field strength is at maximum. At this instant the field is stationary and the voltage output is zero. As the tooth moves away from the magnet, the air gap increases and field strength decreases. This decrease (movement) of the field causes a voltage to be produced in the coil but since field movement direction is reversed, the voltage polarity is reversed in the coil. (d) As the tooth moves to a position as shown in Figure 2-23a, the field strength is weak (stationary) and the voltage output is zero.

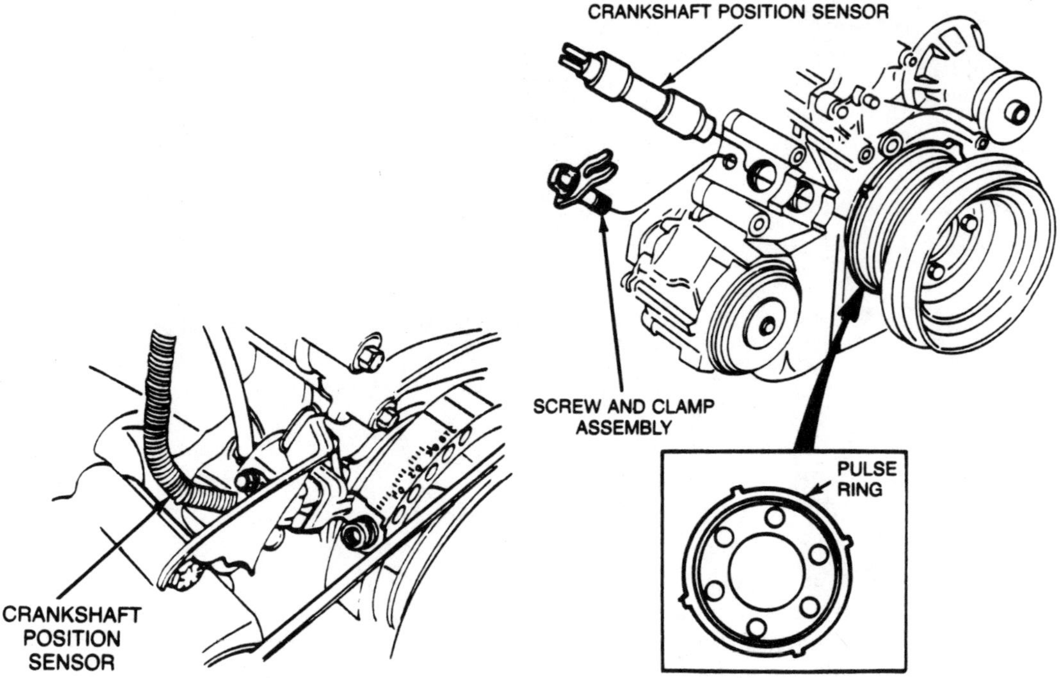

Figure 2-24 Crankshaft position sensor. (Courtesy of Ford Motor Company)

When the Hall effect sensor produces a high voltage signal (no shutter present in air gap) the base of the transistor is conductive, allowing current to flow from the voltage supply line to the collector, the emitter, and then to ground. This result occurs with the signal line reading only a one-volt drop (see Figure 2-27a). When the Hall effect sensor produces a low voltage signal (shutter present in air gap) the base of the switching transistor is no longer conductive, turning the transistor off. With no ground for the voltage supply line the signal line reads 12 volts (see Figure 2-27b). This digital signal (also called square wave signal) is used by the computer to determine engine RPM's and crankshaft position.

The number of shutters equals the number of cylinders and their position is determined by dividing the number of cylinders into 360 degrees (except odd-firing engines). On some engines, particularly those with sequential fuel injection, it is important for the computer to know where number one cylinder piston position is. The shutter that corresponds to number one cylinder will be a dif-

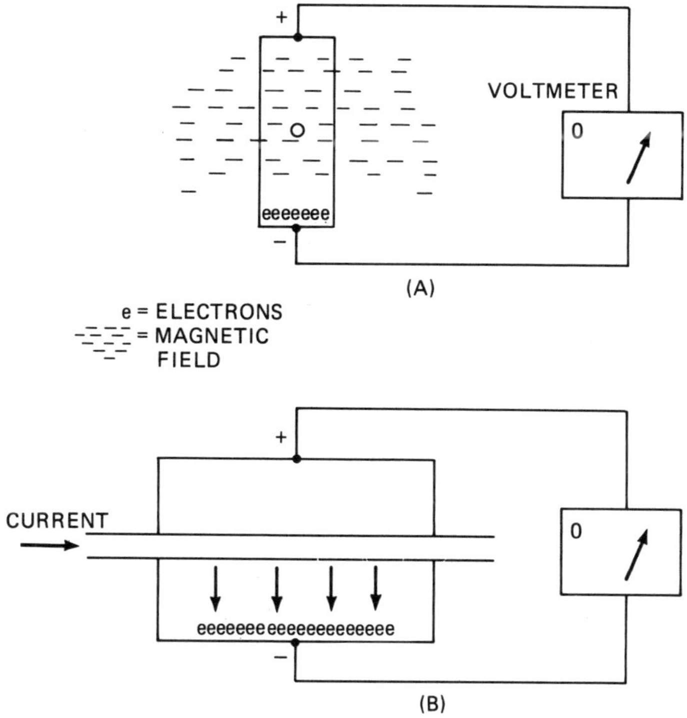

Figure 2-25 (a) Hall element, end view. (b) Hall element, side view.

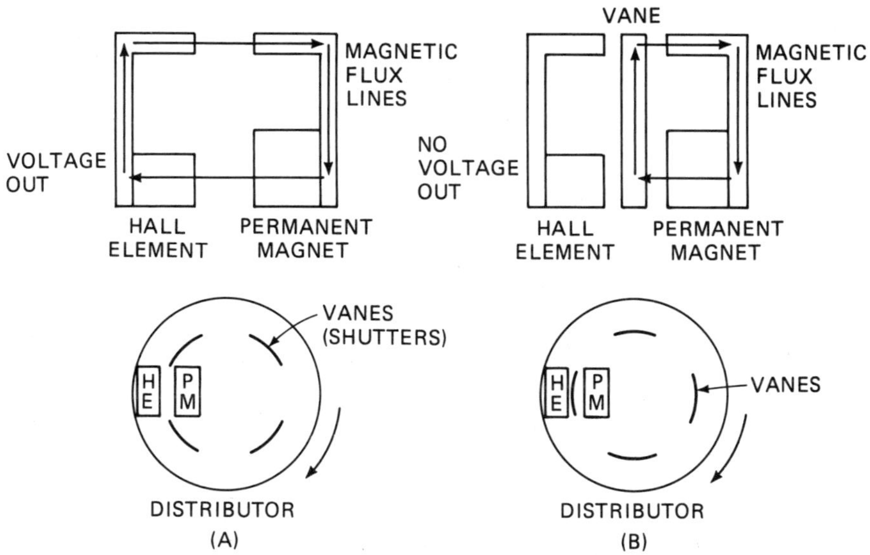

Figure 2-26 Hall effect sensor. (a) No shutter present. (b) Shutter present.

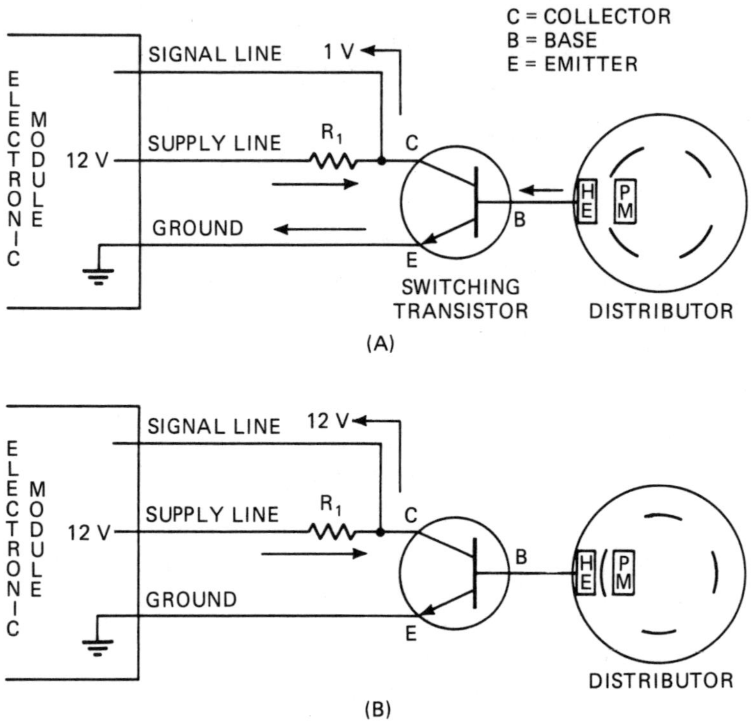

Figure 2-27 Hall effect sensor. (a) High voltage signal. Since R1 has most of the resistance in the circuit from the 12 volt supply to ground, most of the voltage will be used up (dropped) across R1. The signal line will see the remaining voltage on one volt. (b) Low voltage signal. With no voltage signal on the base of the switching transistor, the transistor is switched off (circuit from collector to emitter is open). Since no current is flowing in the supply line, there is no voltage used up (dropped) across R1.

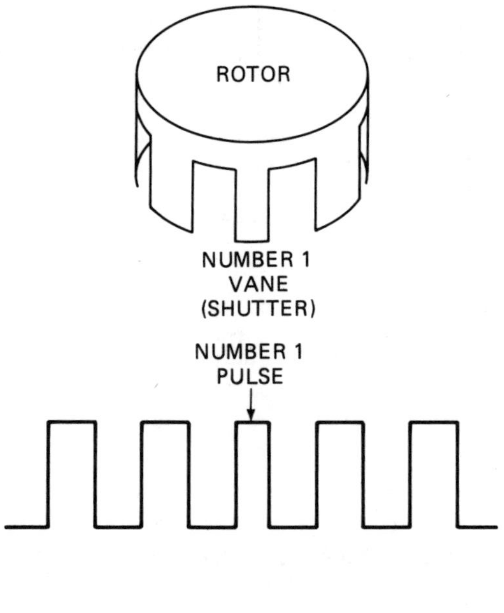

Figure 2-28 Different signal indicates position of number-one cylinder.

ferent width from the rest, which in turn will produce a Hall effect signal of a different duration (see Figure 2-28). This different signal will indicate to the computer cylinder number one piston position.

2-2-7 Digital Switches

Digital switches are on/off devices. That is, in one mode they are on, and in the other mode they are off. A common digital switch you are used to is the light switch, it is either on or off. On cars, digital switches are used to tell the computer whether a device is on or off. They are found in brake switches, neutral safety switches, clutch engagement switches, and any other application where the computer has to know whether a system is on or off. Typically, these switches are open when unloaded. The computer sends a 12 volt (may be 5v or 8v in different applications) signal to the switch, but since the switch is open the computer sees a 12 volt drop. When the switch is closed (stepping on the brake pedal for example) current will flow through the switch and the computer will see a voltage signal less than one. In this manner the computer knows whether a circuit or system is on or off (see Figure 2-29).

2-2-8 Knock (Detonation) Sensors

A knock sensor is a tuned piezoelectric accelerometer which is designed to signal an ignition module or computer that spark knock (detonation) is occurring. Piezoelectricity literally means pressure electricity. That is, when certain crystals, such as quartz, are sub-

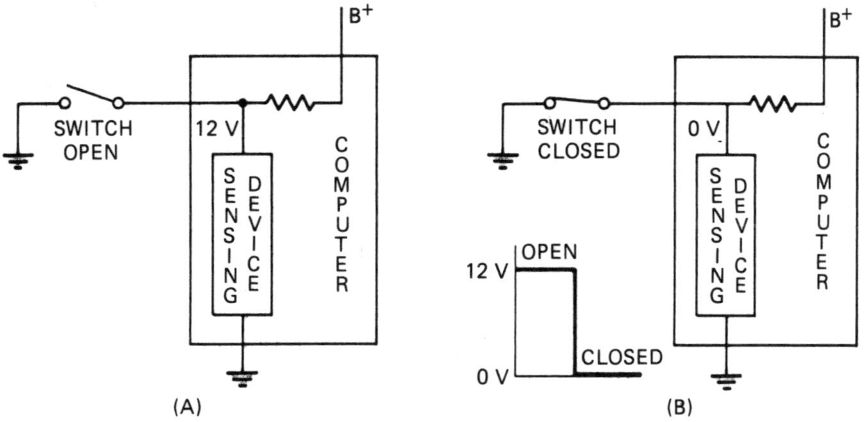

Figure 2-29 Digital switches. (a) Open. (b) Closed.

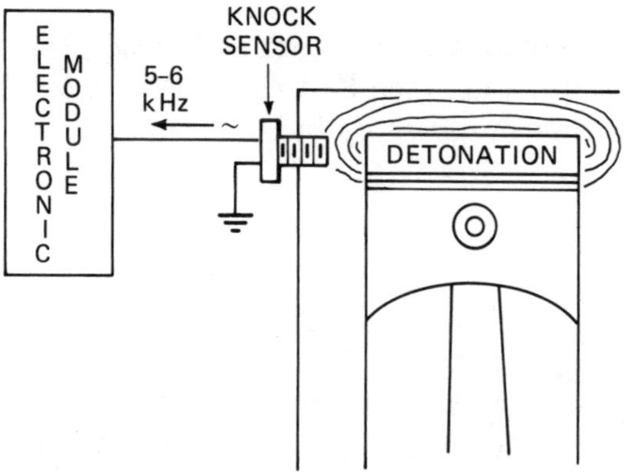

Figure 2-30 Knock sensor.

jected to pressure, voltage is generated. An accelerometer measures acceleration. Vibration is oscillating acceleration. When spark knock occurs, a vibration between 5kHz and 6kHz is emitted. The crystal inside the knock sensor is tuned specifically for this frequency. The vibrations from knock apply pressure on the crystal which in turn generates a voltage. This AC voltage signal is less than one volt but the frequency is from 5kHz to 6kHz (see Figure 2-30). When the computer reads this signal it will retard spark timing until knock disappears and therefore the knock sensor will not generate a signal. The computer may then advance timing back to the optimum or until knock occurs again.

2-2-9 Photoelectric (Phototransistor) Sensors

Photoelectric sensors are typically used to measure vehicle speed. They consist of a light emitting diode (LED), a phototransistor, and a reflective blade or disc. When light from the LED reaches the phototransistor, which is sensitive to light, it will switch on. When the reflective blade or disc prevents light from reaching the phototransistor it will turn off (see Figure 2-31). This digital signal is sent to the computer, where the number of pulses is counted over

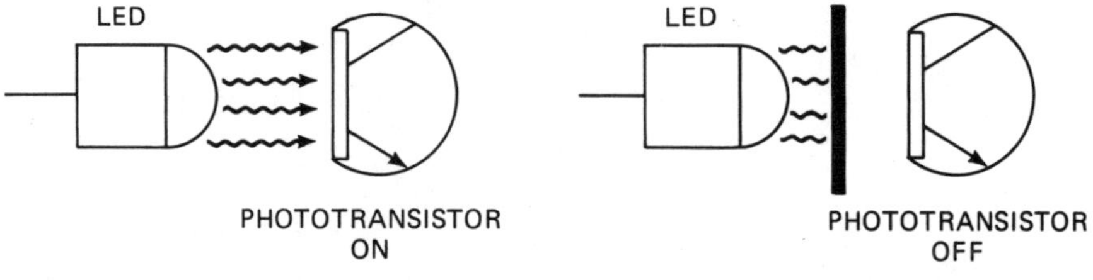

Figure 2-31 Photoelectric sensor. Phototransistor (a) on and (b) off.

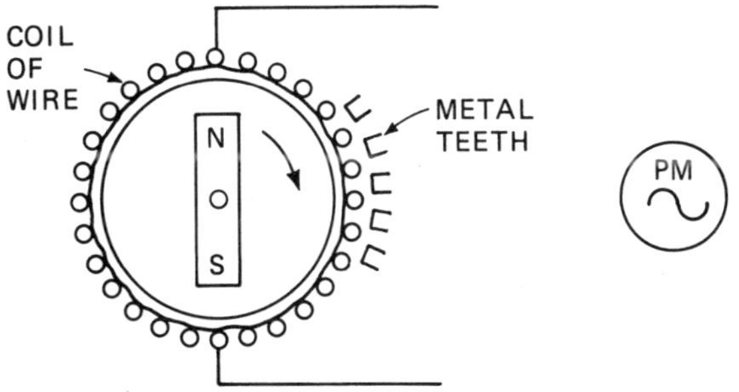

Figure 2-32 Permanent magnet generator.

a given period. When the reflective blade is mounted on the speedometer cable, the number of pulses will indicate vehicle speed.

2-2-10 Permanent Magnet (PM) Generators

PM generators are AC voltage generators which are also used to measure vehicle speed. A small permanent magnet is located inside a coil of wire. As the permanent magnet is rotated the magnetic lines of force induce an AC voltage (see Figure 2-32). The frequency and strength of the induced AC voltage increase as the speed of the rotating magnet increases. This AC voltage signal is converted into a digital signal the computer will understand.

2-2-11 Mass Air Flow (MAF) Sensor

The mass air flow sensor, located in the intake air stream, measures the quantity (mass) of air taken into the engine. There are different types of MAF sensors.

One type uses a heated wire placed in the intake air stream (see Figure 2-33). As air passes around the wire it will tend to cool off; the more air that passes by, the greater the cooling. Additional current is needed to maintain the required resistance in the wire, and it is this additional current which is used to signal the computer about mass air flow. The computer is able to convert the signal and calculate air flow in grams per second. After certain conditions are met and the ignition key is turned off, current will be sent through the wire, heating it red hot. This will burn off any residue that may have coated the wire.

Another type consists of an air temperature sending resistor, heated sensing element, and an electronic module (see Figure 2-34). The air temperature sending resistor measures incoming air

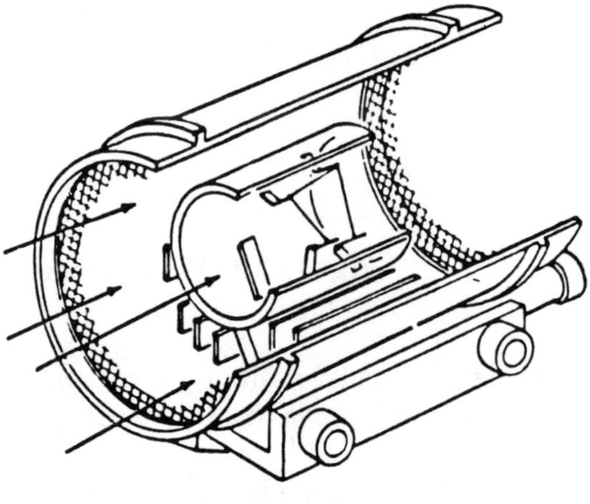

Figure 2-33 Mass air flow sensor—hot wire type. (Courtesy of General Motors Product Service Training

temperature. The sensing element always tries to stay 75°C above incoming air temperature. As the passing air tries to cool the sensing element down, extra current is supplied to the sensing element. As more air passes by the sensing element, more current is needed to keep it 75°C above incoming air temperature. This amount of extra current is used to measure air flow. This extra current is con-

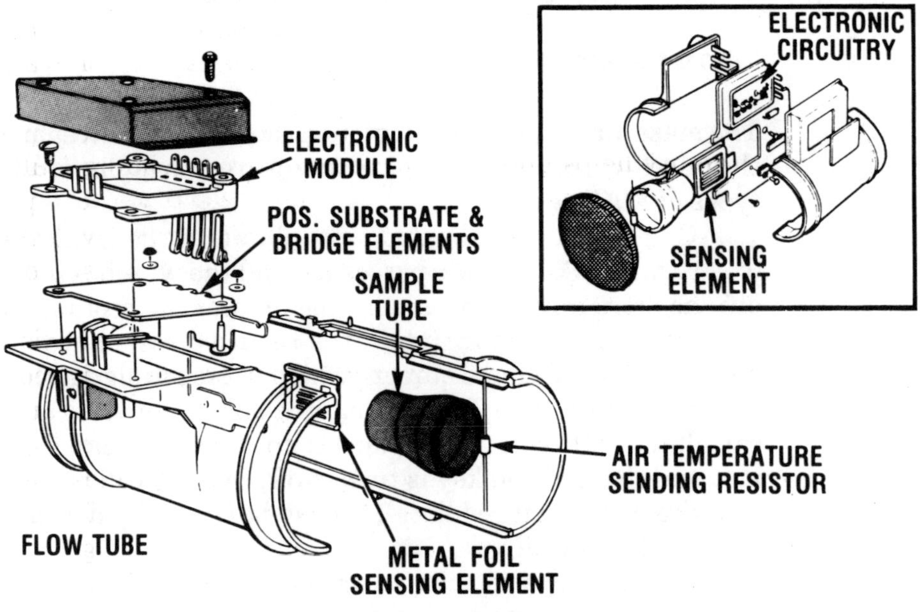

Figure 2-34 Mass air flow sensor—hot film (frequency type). (Courtesy of General Motors Product Service Training.)

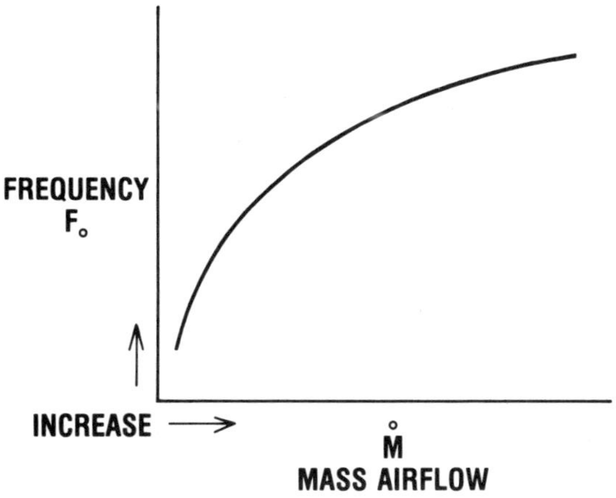

Figure 2-35 As air flow increases frequency increases. (Courtesy of General Motors Product Service Training)

verted to specific frequency within the MAF electronic module. The greater the extra current needed (and therefore the higher the mass air flow), the higher the frequency (see Figure 2-35).

2-2-12 Input Circuit Wires and Connectors

In most sensor circuits a five volt reference source is used. The reason five volts is used is because under all operational modes the system should always have five volts available. For example, when the engine is cranking, battery voltage may go as low as 9.5 volts, but this is still well above the five volts in the input circuits. This stability helps prevent the computer from reading faulty information. Another reason is that most electronic components are designed to run on five volts. But because only five volts are used with very high resistances, the signal lines are easily disturbed by distortion and noise.

Signals to and from the computer must be free of distortion and noise. To reduce distortion, the connectors are designed to prevent the corrosive attacks of the environment. They keep out water, oil, dirt, and other foreign materials that would corrode the contact areas. Corrosion would cause a change in circuit resistance further changing the voltage signal. The computer would then receive an incorrect signal and deliver faulty decisions (see Figure 2-36). Noise in the input circuits may be caused by electromagnetic radiation. The static you hear on the radio interferes with what you want to listen to. A noisy signal line interferes with the normal signals and

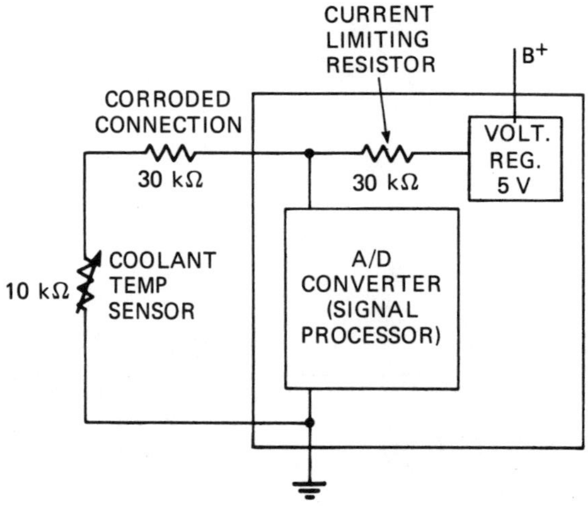

$$R_t = 10\ k\Omega + 30\ k\Omega + 30\ k\Omega$$

$$R_t = 70\ k$$

$$I = \frac{5\ V}{70\ k}$$

$$I = .000071\ A$$

Figure 2-36 A corroded connection adds resistance. The computer wll think the engine is colder than it really is. Using Ohm's law, you can calculate the difference. The voltage drop across the corroded connection is .00071a x 30kΩ = 2.14v. The voltage drop across the coolant temperature sensor is .000071a x 10kΩ = .71v. The A/D signal processor will read the combined voltage drop of 2.14v + .71v = 2.85v. If you compare this to figure 2-4b6, there is a 1.6v difference, and this is significant.

the computer may misinterpret the signal and deliver a faulty decision. To prevent noise, special shields and shielded cables may be used. Also, the wiring harness is routed in a manner that will shield it from sources of interference.

§ 2-3 THE COMPUTER

The computer on today's automobiles performs a variety of tasks, and it is called different names by manufacturers. Specifically, a computer is a set of electronic circuits and components that receives signals, stores information, performs calculations, makes decisions according to a set of instructions, and sends out the appropriate signals. Inside the computer housing, there are other components to help the computer to operate such as voltage regulators and current limiting resistors. For these reasons, manufacturers may call the computer and its components within the housing an electronic control module or electronic control assembly. There are other names, but they are common in their function and purpose. This section is

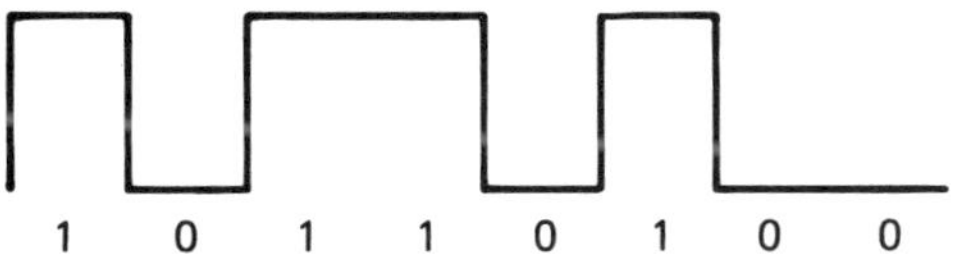

Figure 2-37 One byte. The digital system transmits information by a series of timed electrical pulses. This graph represents eight pulses (bits).

divided into two parts, the first on basic computer operation, the second on other components found within the computer housing.

2-3-1 Basic Computer Operation

Computers operate on digital binary signals. A high voltage will represent a one and a low voltage will represent a zero. In the binary number system you have zero and one. The number two of the binary system would be represented as 10, number three as 11. Each space occupied by a number is called a bit. It takes eight bits to equal one byte or word. This byte would be used to represent a command or piece of information (see Figure 2-37). The components where bytes are created and stored are called integrated circuits (IC) or chips because of their shape, Figure 2-38. The number of bytes the IC chips and computer can hold is termed, for example, 64 kilobytes or 256 kilobytes.

The IC chip that performs the calculations and makes decisions is called the microprocessor. The microprocessor can be an 8 bit, 16 bit, or higher. The higher the number, the faster calculations can be performed.

Information to the microprocessor comes from other types of IC chips, generally called memory circuits. A memory circuit designed for permanent storage of information is called a Read Only Memory (ROM) chip. Information is put in the ROM chip from the factory and it will not accept any new information. The ROM chips provide information to the microprocessor and are considered permanently

Figure 2-38 An IC chip.

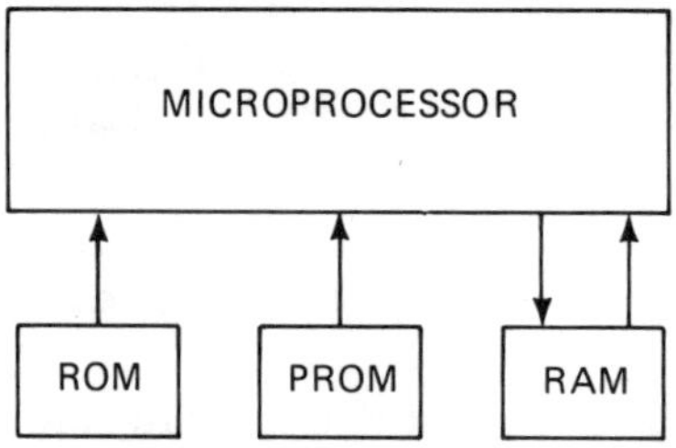

Figure 2-39 Computer memory circuits.

attached to the circuit board. Another type of ROM is called the PROM which stands for Programmable Read Only Memory. The PROM is essentially like the ROM except that it can be customized for different applications.

Random access memory (RAM) chips can store new information and have their memory rewritten as determined by the microprocessor (see Figure 2-39). There are two types of RAM: volatile RAM and nonvolatile RAM. Volatile RAM will lose its memory if it loses its power source. Nonvolatile RAM will retain its memory if removed from a power source.

Another essential part needed for every computer is a clock. The clock provides an accurate and steady pulse. This is necessary for the computer to determine the length of a word and when the pauses occur between words (see Figure 2-40).

2-3-2 Additional Components Within the Computer Housing

The memories, the microprocessor (s), and clock are the basic components of a computer, but the computer has to respond to its in-

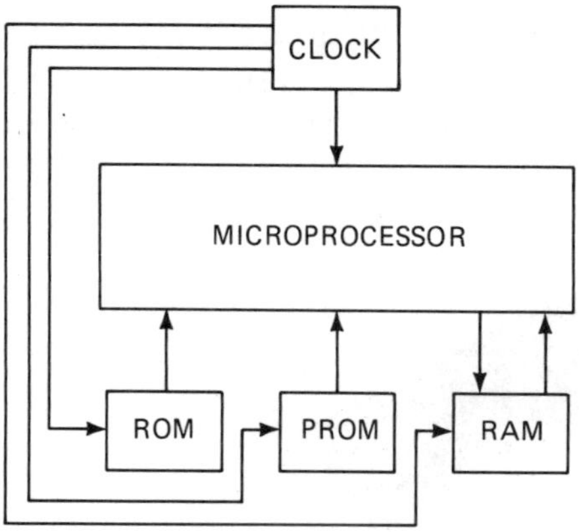

Figure 2-40 Every computer needs a clock.

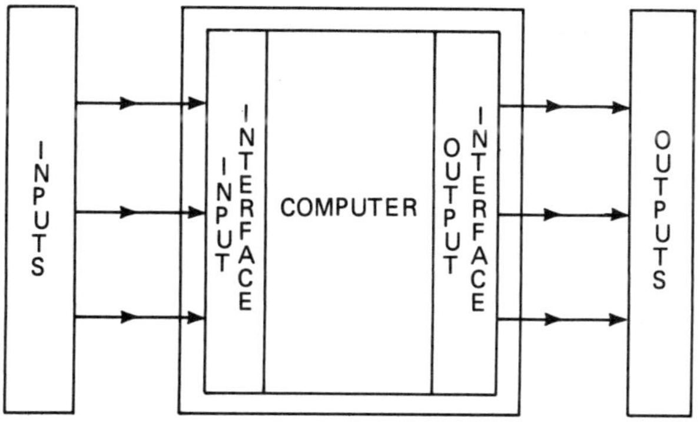

Figure 2-41 Computer interfaces.

puts and outputs. These components are generally called the interfaces. There are input interfaces and output interfaces (see Figure 2-41).

A common interface is the analog to digital (A/D) converter (signal processor). This special chip must convert analog signals from variable resistor inputs (potentiometers, thermistors) into digital signals the microprocessor can understand (see Figure 2-42). Current limiting resistors are used in thermistor circuits to provide a means for the A/D signal converter to measure the voltage drop across the sensor (see Figure 2-43).

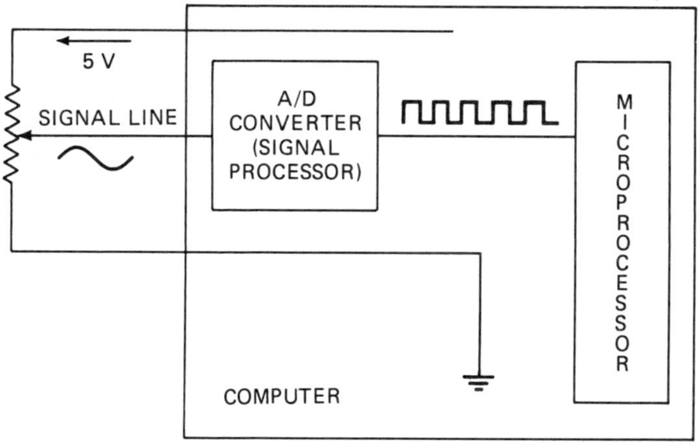

Figure 2-42 The A/D converter (signal processor) changes the analog (variable voltage) signal from a sensor to a digital (on/off) signal that is timed to match the computer frequency.

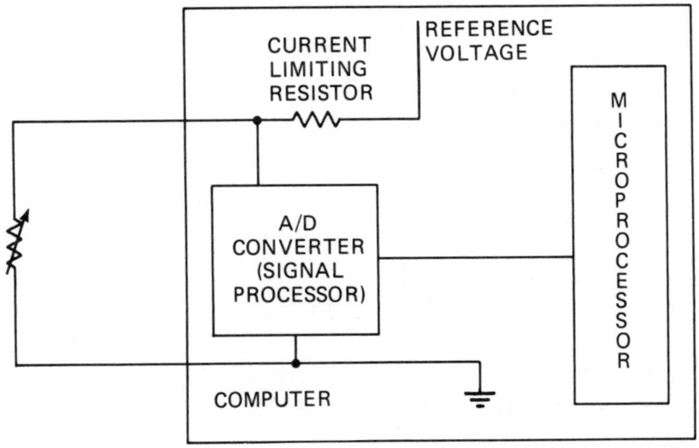

Figure 2-43 Current limiting register in thermistor circuit.

Another interface is the counter. It counts pulses such as those sent by a crankshaft position sensor and sends the count to the microprocessor (see Figure 2-44).

A buffer converts an AC signal into a digital square wave signal. It does not send a count like the counter. The main component is a transistor which will switch on or off according to the polarity of the AC signal (see Figure 2-45).

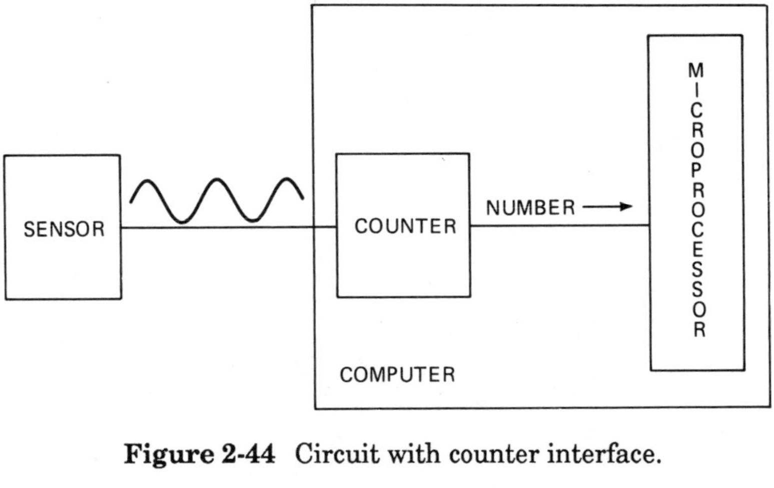

Figure 2-44 Circuit with counter interface.

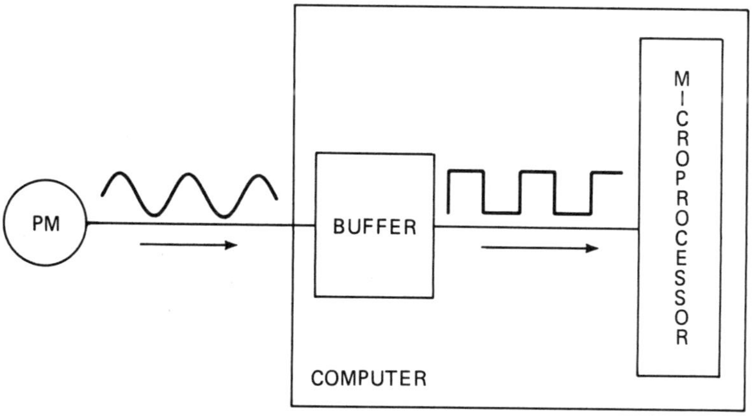

Figure 2-45 This buffer changes the AC signal from the permanent magnet generator to a digital signal.

An amplifier interface modifies a signal from a sensor and sends an amplified signal to the microprocessor for processing, (see Figure 2-46).

A voltage regulator is used to drop battery voltage down to five volts. This helps maintain accurate input signals (see Figure 2-47).

Output interfaces, controlled by the microprocessor, consist of switching transistors (also called output drivers). When the microprocessor decides to turn on a device, it will send a very tiny current to the base of the appropriate transistor which will in turn

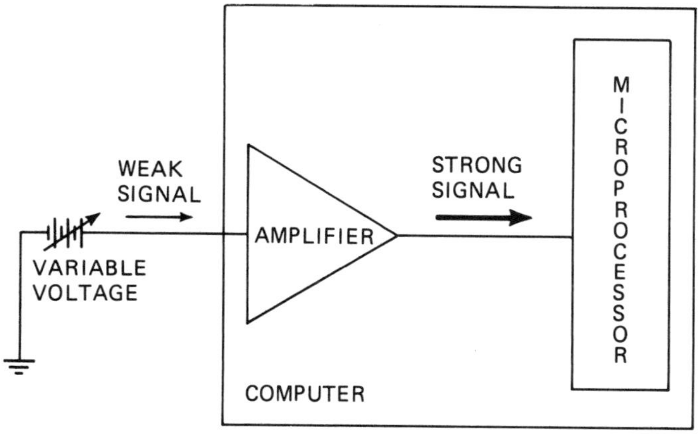

Figure 2-46 Amplifier interface.

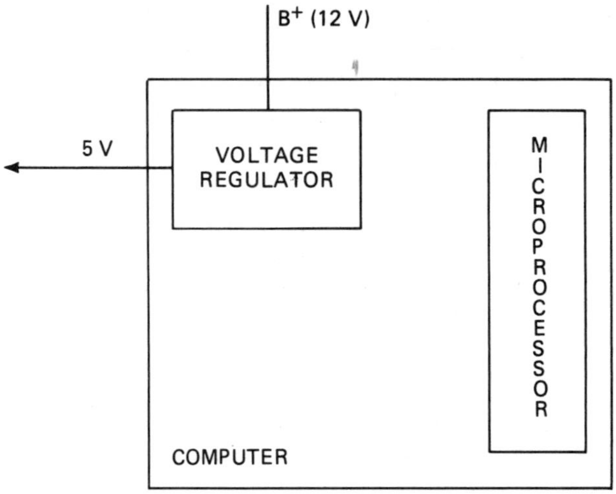

Figure 2-47 Voltage regulator.

complete the circuit, allowing the device to turn on. To turn a device off, the computer will remove the current applied to the transistor base and the transistor will turn off, opening the circuit (see Figure 2-48). The output devices the computer controls are discussed in the next section.

§ 2-4 OUTPUTS

Outputs are controlled by the computer to modify operation or send information. Outputs can be divided into four categories: solenoids, relays, motors, and readouts.

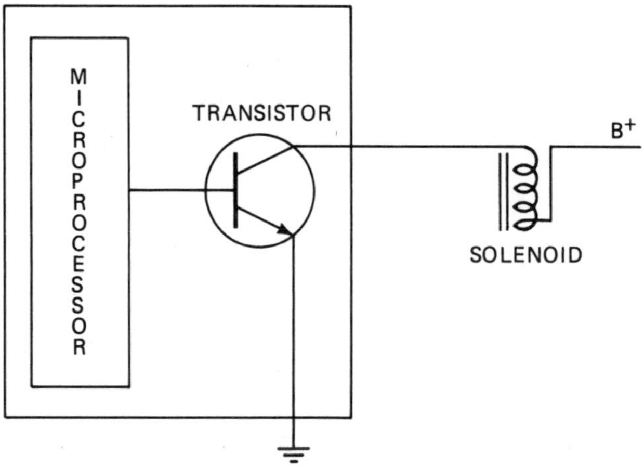

Figure 2-48 Typical output interface or driver.

2-4-1 Solenoids

A basic solenoid consists of a coil of wire surrounding a spring-loaded plunger. When current passes through the coil the magnetic field forces the plunger against the spring. Though there are many solenoids used in a variety of applications they fall into two basic categories, normally closed solenoids, and normally open solenoids. A normally closed solenoid will not allow the passage of any vacuum, fluid, or air when the solenoid is not energized. Energize the solenoid and the plunger will retract, allowing flow through the valve and seat area (see Figure 2-49). Fuel injection nozzles and nearly all vacuum control solenoids are normally closed. A normally open solenoid permits flow through the solenoid until it is energized, then flow stops (see Figure 2-50).

2-4-2 Relays

Relays are used in applications where a small current flow is used to control a large current flow. The small current flow portion is the control circuit, the large current flow portion is the load circuit (see Figure 2-51). Relays make it possible for the computer to control high load devices such as the air-conditioning compressor, radiator-cooling fan, or fuel pump. The computer activates the control circuit which energizes the relay. When the contacts close in the relay, current flows through the load circuit activating whatever device

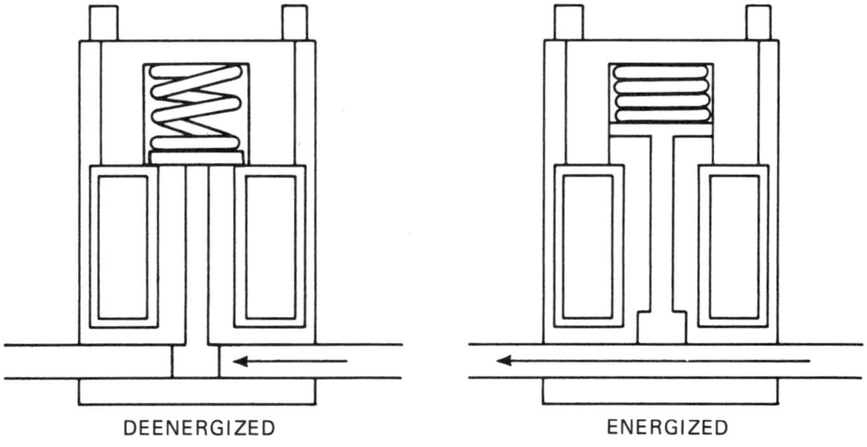

Figure 2-49 Normally closed solenoid.

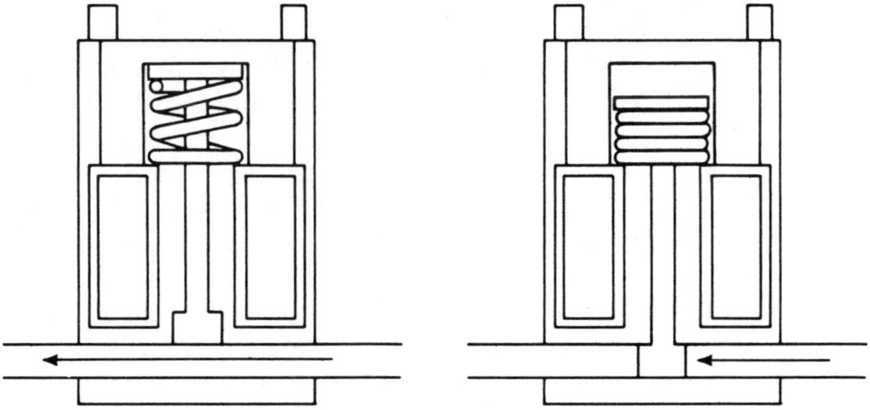

Figure 2-50 Normally open solenoid.

the relay was connected to. Use of the relay permits small amounts of current easily handled by the computer to control large current flow. It would be impractical and less efficient to use high current wires from the computer to the load in most cases. First, the transistors (output drivers) would have to be large to accommodate the load. Second, heavy cables would have to be strung a greater distance which would increase resistance and make connections to the computer difficult.

2-4-3 Motors

Electric, reversing DC motors are commonly controlled by the computer. They are often used to control idle speed by moving the throttle valve. The motor will have a plunger that will extend or retract according to direction of rotation. The direction of rotation is determined by the computer. For example, as loads increase on the engine and idle speed drops, the computer will energize the circuit which will cause the plunger to extend, raising idle speed.

A DC stepper motor is also used in many cases. This motor rotates its armature and extends or retracts a plunger a fixed amount. This fixed amount is called a step. The computer controls the field windings and will energize the proper field windings to move the armature in the proper direction a fixed amount. These motors are often used to manage air or fuel flow (see Figure 2-52).

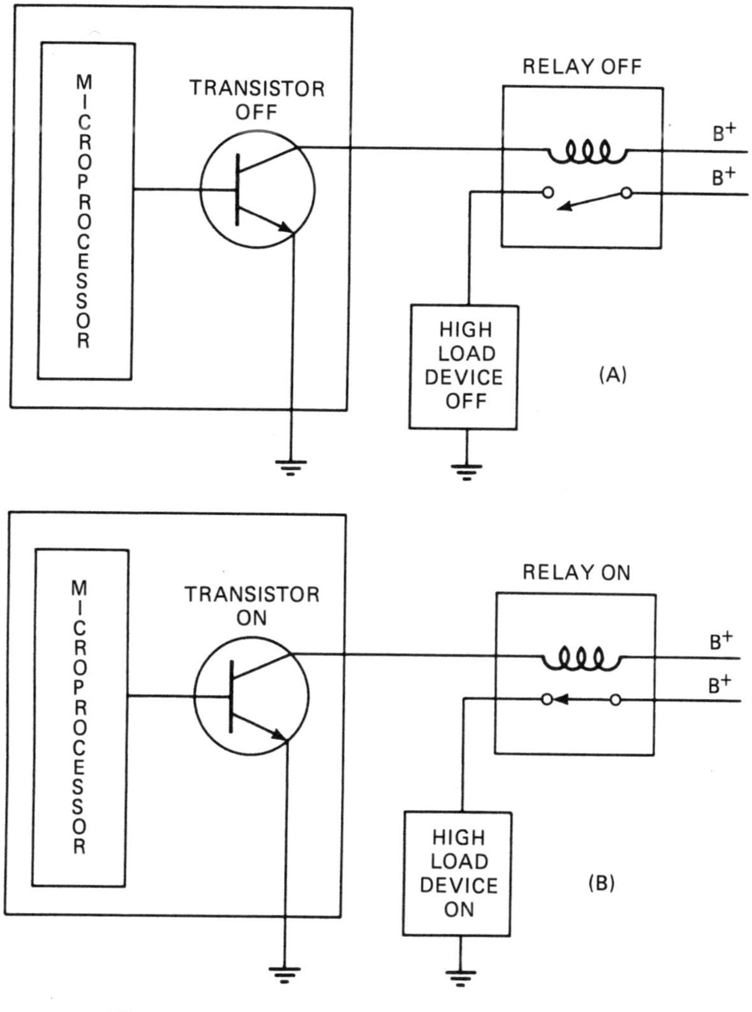

Figure 2-51 Relays shown (a) off and (b) on.

2-4-4 Readouts

Readouts are devices used by the computer to provide information to the driver. A readout can be as simple as a light bulb turning on to indicate trouble or as complicated as digital instruments (digital instruments are discussed in Chapter 4). When a technician wishes to investigate a computer circuit and computer problems, the computer will provide a readout in the form of number codes after the proper steps are initiated, and will indicate if there is a problem.

2-4-5 Ground Side Switching

Regardless of manufacturer, nearly all outputs are activated by what is known as ground side switching. That is, the switch to activate the output is located on the ground side of the output. When the computer decides to activate a device, the transistor (output driver) turns on, completing the circuit path to ground (see Figure 2-53). With ground side switching, any short to ground will occur before the computer and prevent computer damage (see Figure 2-54). In some instances, the computer may be located on the power (hot) side of the output. This is done either for convenience or because it may be better to have the computer damaged than risk safety or damage to more expensive components.

2-4-6 Output Circuit Protection Devices

The computer must be protected from high voltage surges called voltage spikes. A voltage spike occurs whenever a device that contains a coil of wire, such as the air-conditioning compressor motor, cooling fans, or solenoids, for example, are shut off. When current

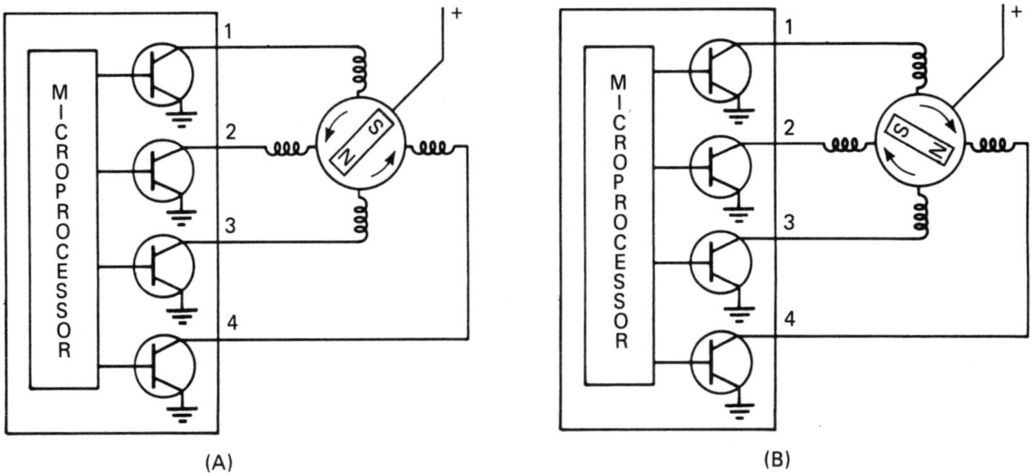

Figure 2-52 Stepper motor functions. (a) Computer grounds 1, then 2, then 3, and then 4. The rotor turns counterclockwise. (b) Computer grounds 4, then 3, then 2, then 1. The rotor turns clockwise.

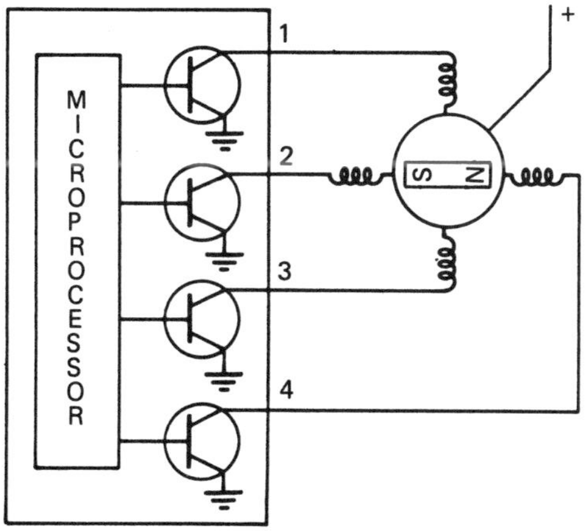

(C)

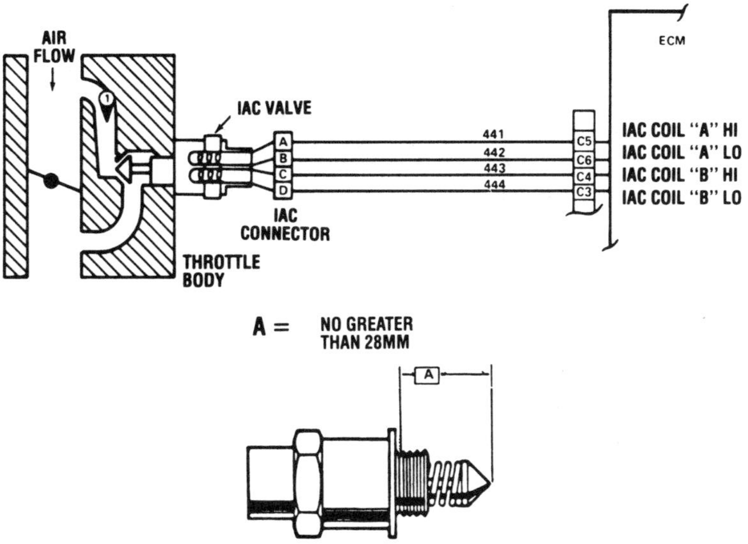

Figure 2-52 (cont.) (c) Computer grounds 1, then 2, and stops. The rotor is now aligned with coils 2 and 4. (d) Stepper motor in idle air passage. (Courtesy of General Motors Product Service Training)

flow is cut off, the magnetic field surrounding the coil of wire collapses, inducing a high voltage. This voltage may feed back to and damage the computer. Diodes, sometimes called clamping diodes, are connected in parallel across any device containing a coil that can induce a high voltage spike (see Figure 2-55). The diode, providing

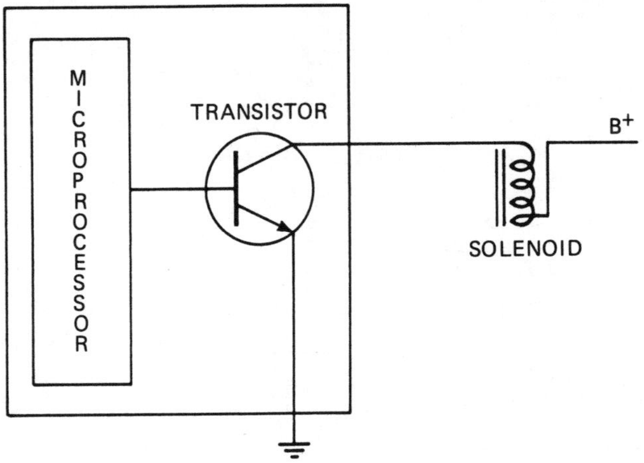

Figure 2-53 Solenoid is not activated until the transistor is turned on, completing the path to ground.

an alternate path, reduces voltage pressure and prevents computer damage. Another device is a high ohm value resistor connected in parallel across coils, solenoids, and so forth. This provides the same protection (see Figure 2-56).

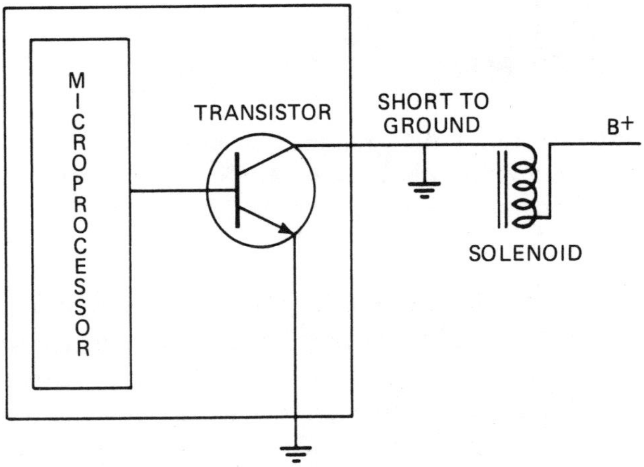

Figure 2-54 Solenoid is activated when grounded before the computer. The computer, however, is unharmed.

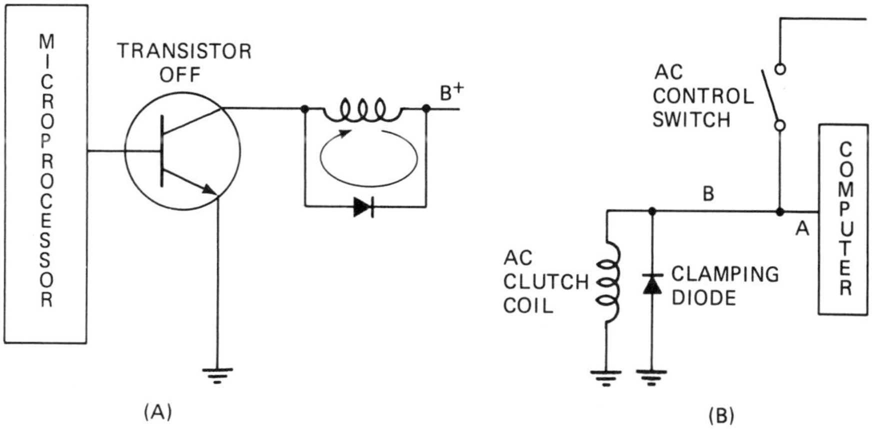

Figure 2-55 Clamping diode. (a) Circuit on. (b) Circuit off. During normal operation (circuit on) the diode is connected with a polarity that will not allow it to conduct. When the circuit is turned off, the collapse of the magnetic field across the coil induces a voltage of reversed polarity. The diode can now conduct which will allow the energy in the coil to dissipate without causing a voltage spike.

With the AC switch closed, current flows in circuit B which energizes the AC clutch. Circuit A tells the computer the AC is on so the computer can increase idle speed to prevent engine stalling from compressor load. The diode will not conduct current as shown.

When the AC control switch is turned off, the magnetic field around the coil collapses. This collapse generates a voltage in the clutch coil which has an opposite polarity (top, negative; bottom, positive). This polarity is correct for the diode to conduct, so current will flow through the diode. This will prevent voltage from building on circuits B and A and prevent high voltage from damaging the computer.

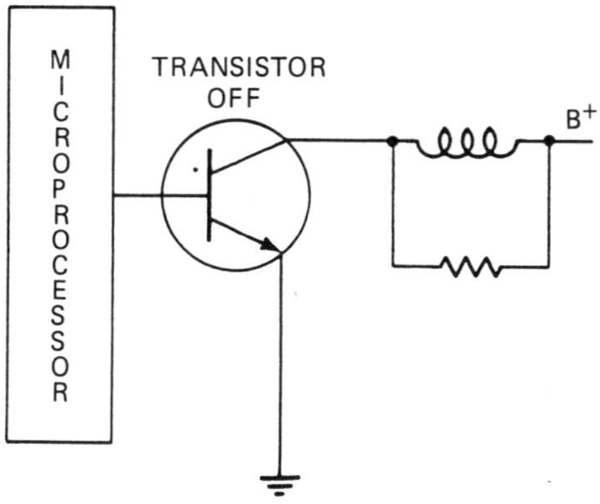

Figure 2-56 A resistor may be used to serve the same purpose as the diode. The resistance is very high (around 600 ohms). The resistor provides an alternative path and prevents a high-voltage buildup which could damage the computer.

QUESTIONS

2-1. What are thermistors and how do they work?

2-2. What is a potentiometer? Describe how it works.

2-3. How does the exhaust gas sensor work?

2-4. Name the three types of MAP sensors and describe how they work.

2-5. Describe how a magnetic reluctance type sensor works.

2-6. Describe the Hall effect. Describe how the Hall effect is used to measure engine RPM.

2-7. Describe digital switch circuit operation.

2-8. Describe how the knock sensor works.

2-9. Describe how the photoelectric sensor works.

2-10. Describe how the PM generator works.

2-11. Describe how both types of MAF sensors work.

2-12. What is a byte?

2-13. What is a microprocessor?

2-14. What do ROM, PROM, RAM, and NVRAM stand for? What are the differences?

2-15. What does an A/D converter do?

2-16. What is the difference between a counter and a buffer?

2-17. What does an amplifier do?

2-18. What is the difference between a normally closed and a normally open solenoid? Which is more common?

2-19. Why are relays used?

2-20. How does a stepper motor work?

2-21. What is ground side switching?

2-22. What is the purpose of a clamping diode?

Chapter 3

Engine Computer Systems and Testing

OBJECTIVES

After studying this chapter the student will be able to

- describe open and closed loop operation
- describe electronic advance operation
- describe the difference between feedback and nonfeedback carburetors
- describe the main difference between throttle body and port fuel injection systems
- describe the main difference between port fuel injection and sequential fuel injection systems
- explain why a digital volt-ohm meter is used in testing computer circuits
- list the precautions when testing the computer and related circuits

§ 3-1 PURPOSE OF ELECTRONIC ENGINE CONTROL

In order to gain optimum performance from a gasoline engine in all its different operating modes (idle, acceleration, cruise, and so forth), it is necessary to make adjustments to various parts of the system. Take ignition timing for example. On early automobiles, drivers had a lever to adjust timing to what they felt was best for certain conditions. Later on, mechanical devices were incorporated (vacuum and mechanical advance), to give more precise control.

At the same time the carburetor systems were refined to provide proper mixtures in various operating modes. As emission controls became necessary, mechanical devices were provided to control exhaust gas recirculation, secondary air injection, canister purge, and so on.

These various devices all operated independently and in some cases their functions interfered with each other.

In modern gasoline engines with electronic controls, all these functions are governed by a common set of sensors and operated by a central control (called a microprocessor, or computer). These functions under one control can now be made to complement each other, and the overall operation of the system provides greatly improved performance, fuel economy, and emission control.

This chapter describes how electronic engine controls manage fuel, ignition, and emission components to provide improved fuel economy, better engine performance, and emission reduction.

§ 3-2 CLOSED LOOP OPERATION

Closed loop operation refers to a system that controls its output by receiving input information from the output. For example, oxygen sensor information is used to adjust the air/fuel mixture. Through the oxygen sensor, the computer is actually able to determine exhaust gas oxygen content and therefore use this extra bit of information to manage the air/fuel ratio. The closed loop mode is achieved when the oxygen sensor has warmed up enough to provide reliable information and the engine is not at idle for extended periods, under rapid acceleration or deceleration, and wide open throttle, Figure 3-1.

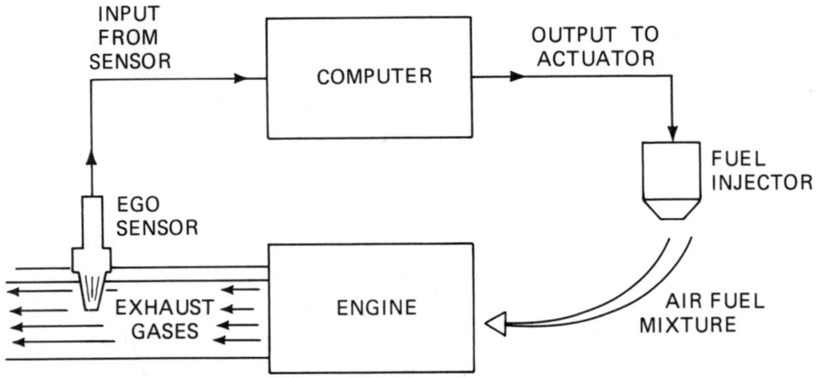

Figure 3-1 How the computer manages air/fuel mixture. If the EGO sensor senses a lean mixture (high oxygen content) its voltage output to the computer will be low (less than .4 volts). The computer will then adjust its output to increase injector pulse width (time open on each pulse), which will add more fuel, producing a richer mixture. The resultant exhaust with a low oxygen content will cause the EGO voltage to be high (more than .6v). This in turn will cause the computer to reduce injector pulse width and make the mixture leaner.

The overall result is a constant cycle from rich to lean and back again which will produce an average air/fuel ratio of approximately 14.7 to 1.

Closed loop is usually obtained only at normal cruise with a warm engine. Some systems will also idle at closed loop under certain conditions.

§ 3-3 OPEN LOOP OPERATION

It is not possible to maintain 14.7 to 1 air/fuel ratio under all conditions. If the engine is cold or under heavy load, richer mixtures are required. Also, if the EGO sensor is not warmed up sufficiently, it is not reliable. In these instances the computer ignores EGO output and controls the mixture by a fixed program which is modified by engine temperature, throttle position, RPM, and load. This mode of operation is called open loop. Feedback from the EGO is not used in computing the air/fuel ratio.

§ 3-4 ELECTRONIC SPARK ADVANCE

As engine load and speed vary under different operating conditions, it is necessary to adjust the ignition timing to provide maximum power output and prevent spark knock (detonation).

The mechanical systems (vacuum and mechanical advance) that were used for many years worked quite well, but their response time to changing conditions was too slow to provide optimum ignition timing most of the time.

In an electronic system, the microprocessor has access (through the sensors) to all of the information necessary to provide correct ignition timing in all situations. Since the system controls electrical signals, the response time is much faster than that of a mechanical system.

The base, or initial, timing signal is provided either by a crankshaft position sensor (a variable reluctance sensor mounted adjacent to a sensing disc attached to the crankshaft), or a Hall effect device, or a variable reluctance sensor mounted in the distributor. The base timing signal is sent to an ignition module or spark advance module. These devices are simply electronic switches to control (switch on and off) the current to the ignition coil primary.

The microprocessor modifies the base timing signal before it gets to the ignition module and total advance is always base timing plus computer timing (see Figure 3-2). The sensors that could affect computer timing include manifold absolute pressure, air flow (vane or mass types), air charge temperature, throttle position, engine coolant temperature, barometric pressure, or knock sensor.

Here is the basic operation. When cranking, no spark advance is applied. Ignition occurs at the base, or reference, timing position. At a predetermined engine RPM spark advance will be applied. The computer calculates the proper amount of spark advance based on the variables mentioned above. A signal is sent to the spark advance or ignition module. The spark advance module acts as a switch in the ignition primary circuit. When the spark advance module receives the signal sent by the computer, it interrupts ignition primary circuit flow, magnetic field around the coil collapses, and current flow is induced in the secondary windings, creating a spark at the spark plug.

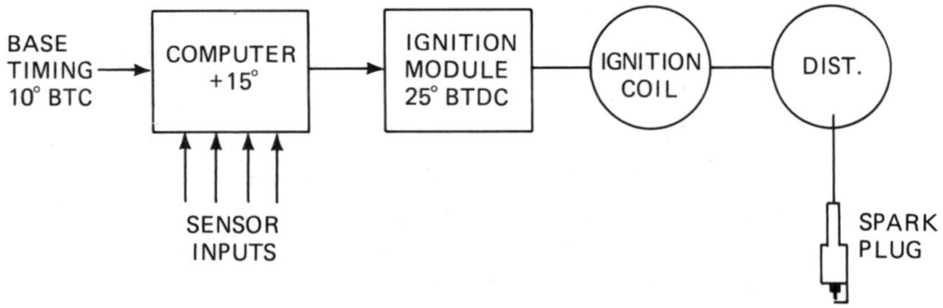

Figure 3-2 How the computer manages ignition timing. In this example base timing is set 10° before top dead center (BTDC). Because of sensor inputs (at this instant) the computer advances the timing an additional 15°. The result is that the ignition module will cause the coil to fire (sent high voltage to the distributor and spark plugs) at 25° BTDC. The total advance then would be 25° BTDC. Total advance is always base timing plus computer timing.

There are periods when spark timing must be decreased. When the knock sensor detects detonation it sends a voltage signal which signals the computer that detonation is occurring. The computer will then calculate timing based on this information and will retard the timing until the detonation stops. After a predetermined period when no knock has been detected the computer will begin to advance the timing back to the optimum level. Should knock occur again the retard process is repeated.

Should the computer fail or signal lines become interrupted, ignition may still occur. However, there will be no spark advance, just base timing only, and the engine will run poorly.

§ 3-5 CATALYTIC CONVERTERS AND AIR MANAGEMENT SYSTEMS

Early catalytic converters made of a platinum/palladium catalyst were designed to convert hydrocarbons (HC) and carbon monoxide (CO) to carbon dioxide and water. In about 1980, another chamber containing rhodium was added to the catalytic converter. This is the front chamber, and it is used to break down oxides of nitrogen (NOX) (see Figure 3-3). But in order for the catalytic converter to do its job efficiently, it must be hot enough and the air/fuel ratio

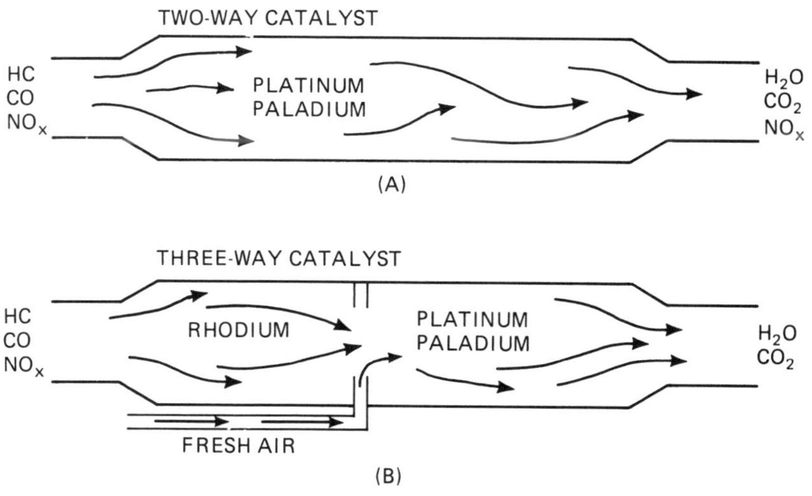

Figure 3-3 Catalytic converters.

must be stoichiometric (14.7 parts air to 1 part fuel). Any richer or leaner air/fuel mixture and converter efficiency is reduced. Electronic control of the air/fuel ratio in closed loop provides the needed control.

Early air pumps simply added fresh air to the exhaust manifold to promote further combustion and reduce HC and CO. On some later models air was added downstream to the catalytic converter to provide a leaner air/fuel ratio and help the converter reduce HC and CO. Computer control of the air management system builds on the earlier systems. When the engine is cold, air provided by the air pump is directed to the exhaust manifold by the computer. This has two purposes. First the extra air in the exhaust manifold helps burn off the rich air/fuel mixture needed for starting, thus lowering emissions. Second, the heat gained by the extra combustion in the exhaust manifold helps to heat the oxygen sensor and the catalytic converter (see Figure 3-4).

As the engine warms up, the computer will direct the air downstream to the chamber of the catalytic converter where HC and CO are converted (see Figure 3-5). The extra air creates a leaner air/fuel ratio, allowing a more complete conversion of HC and CO to carbon dioxide and water.

Should the engine run at fast idle for an extended period, the computer diverts the air to atmosphere. It does this to prevent the catalytic converter from overheating.

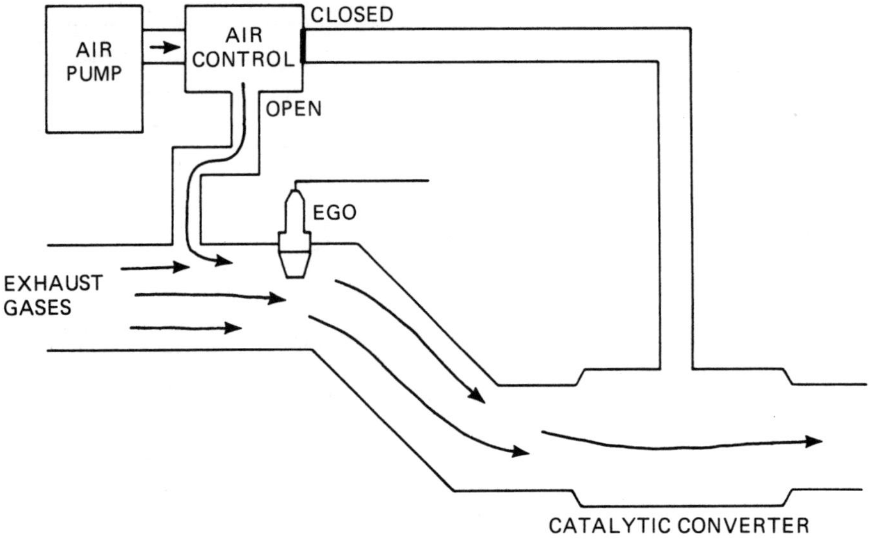

Figure 3-4 Computer control of air management system when engine is cold.

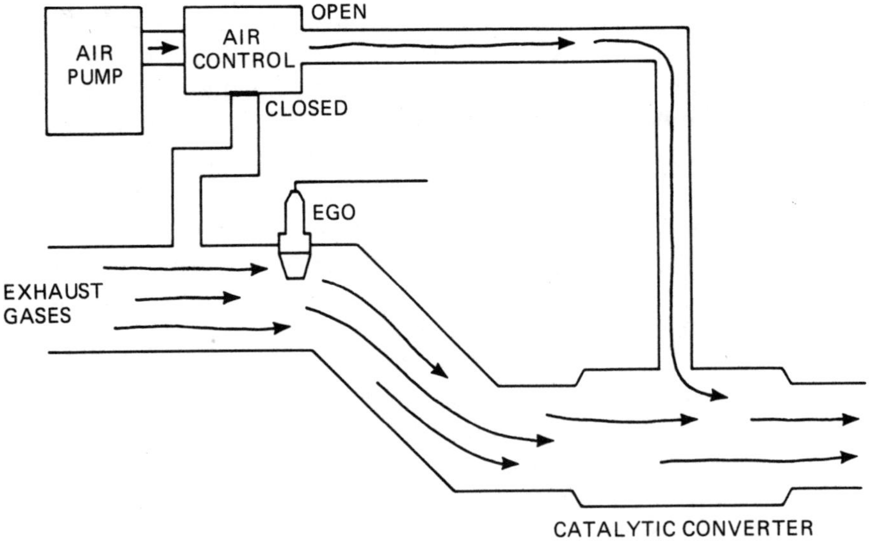

Figure 3-5 Computer control of air management system when engine is warm.

§ 3-6 FEEDBACK CARBURETORS

A feedback carburetor has all the systems of a conventional carburetor. In addition, however, it will have a device that can modify or operate in parallel with the conventional main and idle systems. This device will be controlled by the computer and could be a stepper motor or an electronically controlled vacuum or mechanical fuel control solenoid. The stepper motor, Figure 3-6, could control an air bleed into the main and idle systems. The fuel control (or mixture control) solenoid is cycled from its rich to lean positions ten times per second.

Control of fuel can also be accomplished by a mixture control (MC) solenoid. When the MC solenoid receives a voltage signal from the computer, the MC solenoid restricts fuel flow. With no voltage signal present, fuel flow is unrestricted by the MC solenoid. Specifically, here is how the computer controls the MC solenoid. The computer divides each second into ten equal segments and determines how long the MC solenoid is on in each segment. The on period is referred to as the pulse width. If the MC solenoid is on during half the segment the pulse width would be 50 percent. The richest mixture would result from a pulse width of 0 percent (off all the time). The leanest mixture would be achieved with a 100 percent pulse width (on all the time). The computer controls the MC solenoid when in closed loop, based primarily on oxygen sensor readings (in open loop it will rely on other sensors). Should the oxygen sensor indicate a rich air/fuel mixture, the computer will energize the MC solenoid for a longer pulse width. There will still be ten pulses per second. If a richer mixture is called for, the mixture control solenoid will be energized for a shorter pulse width. Again, there will still be ten pulses per second.

Control of the air/fuel mixture can also be accomplished by controlling the amount of air in the idle and main circuit passages. Again, a solenoid controlled by the computer will determine how much air enters the circuit passages. The more air, the leaner the mixture (see Figure 3-7).

§ 3-7 FUEL INJECTION

Fuel injection offers a more precise method of controlling the amount of fuel delivered to the cylinders. This means that less emis-

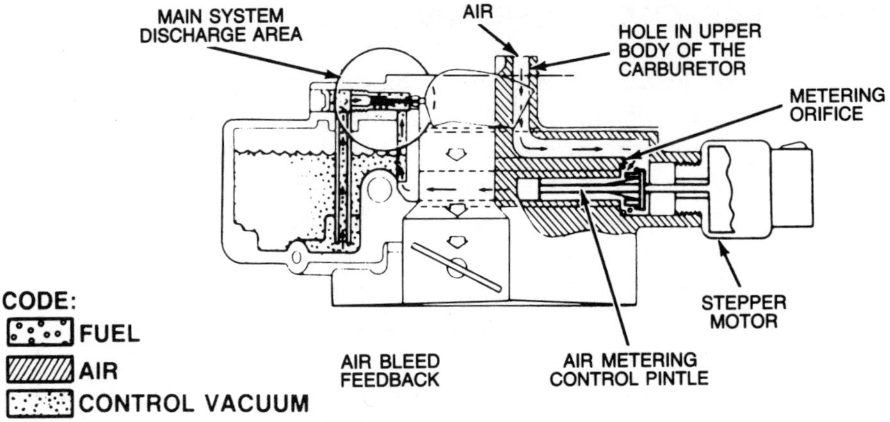

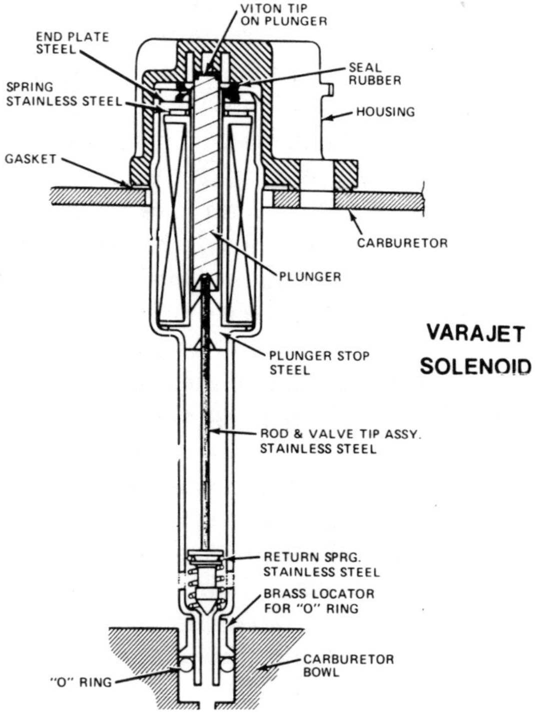

Figure 3-6 Typical mixture control devices. (a) Courtesy of Ford Motor Company. (b) Courtesy of General Motors Product Service Training.

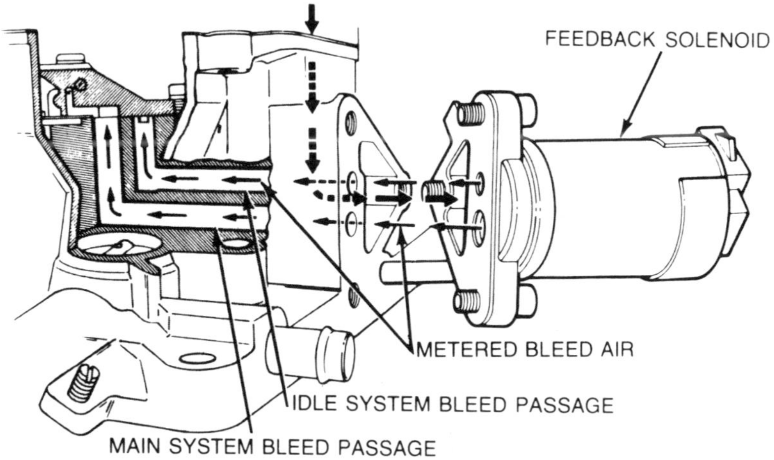

Figure 3-7 Courtesy of Ford Motor Company.

sions, greater economy, and more horsepower can be achieved over a carburetor-equipped engine. A primary disadvantage of fuel injection is a higher initial cost, though this may change in time.

Basically, gasoline fuel injection involves fuel being sprayed into the intake tract at different times, or pulses. The fuel injector is a normally closed solenoid. When energized by the computer, the needle valve is pulled off its seat and fuel under pressure escapes the injector (see Figure 3-8). The location of the fuel injector(s) is what primarily determines what type of injection system it is.

3-7-1 Throttle Body (Central Fuel) Injection

Throttle body, or central fuel, injection system uses one or two fuel injectors to supply fuel to all cylinders. The fuel injector(s) is located above the throttle plate which means the fuel is sprayed out into atmospheric pressure (see Figure 3-9).

A fuel pump located near or in the fuel tank supplies fuel under pressure to the injector. Fuel pressure ranges anywhere from 15 psi to 40 psi depending on the system type. The fuel pressure regulator maintains a predetermined setting. If pressure becomes high, fuel is bled back to the fuel tank through the return line. The regulator is preset and cannot be adjusted, only checked and replaced.

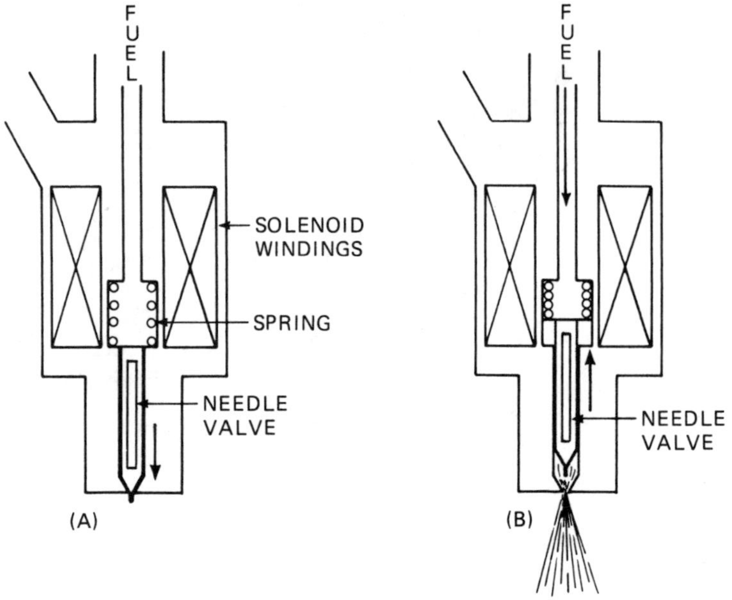

Figure 3-8 Fuel injection system. (a) Injector closed. (b) Injector open.

The amount of time the fuel injector is on is first determined by engine demand. When the computer sees throttle position increasing, it will increase the time or pulse the fuel injector is on. When the engine is cold, injector on-time increases to provide a richer mixture. On rapid deceleration, the fuel injector can be momentarily turned off. This helps to increase fuel economy and lower emissions.

Because the fuel injector is mounted in the same place as a carburetor, it also needs a heated air intake to promote fuel vaporization when the intake tract is cold. The heated air intake operates exactly the same as it would if mounted to a carburetor.

3-7-2 Port Fuel Injection

Port fuel injection uses one fuel injector for each intake tract to a cylinder. Each fuel injector is located in the intact tract just before the intake valve and fuel is sprayed out into intake manifold pressure (see Figure 3-10). This creates an important difference with throttle body fuel injection.

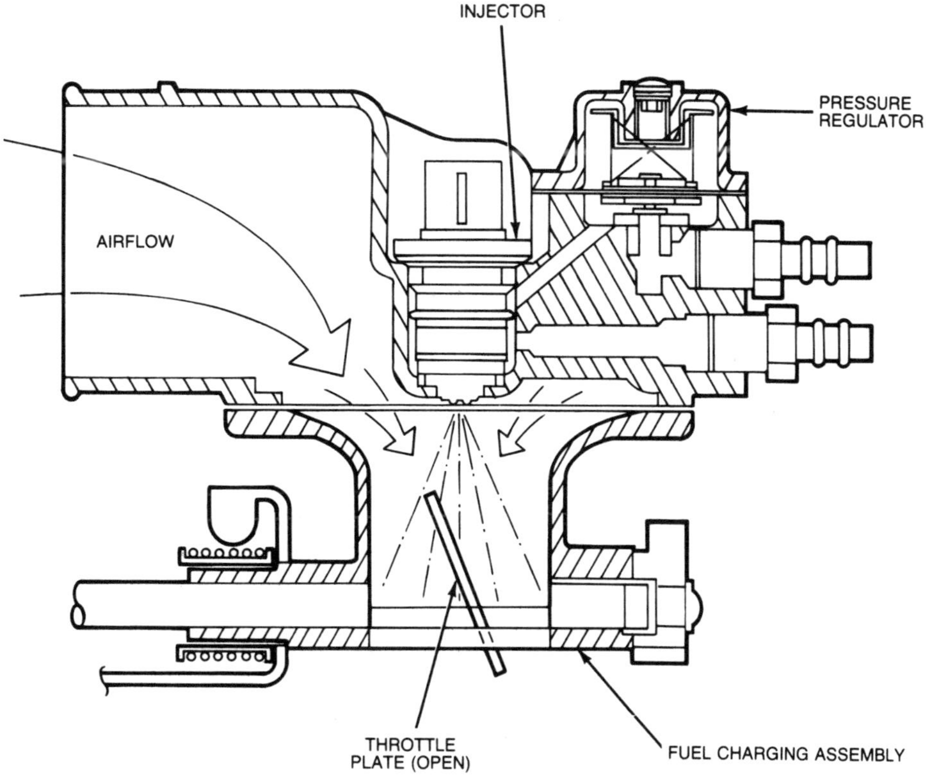

Figure 3-9 Throttle body (central) fuel injection system. (Courtesy of Ford Motor Company)

Fuel pressure is supplied by a fuel pump and regulated by a pressure regulator according to intake manifold pressure (vacuum). The fuel pressure regulator is connected on one side to intake manifold pressure (vacuum) (see Figure 3-11). At idle, when manifold pressure is very low, fuel pressure is also low. At wide open throttle, when manifold pressure is high, fuel pressure is also high. If the fuel pressure did not change with intake manifold pressure (vacuum), too much fuel would be injected at idle and too little fuel at wide open throttle.

With fuel injection it is critical to know the mass of the air entering the engine. The air's mass (or density) is what primarily determines the amount of fuel injected into the cylinder. Two methods are used to find air mass: speed density method or a mass air flow sensor. The speed density method determines air mass based on

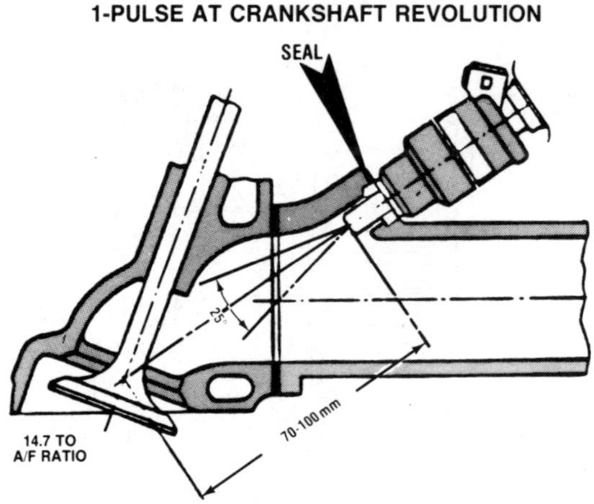

Figure 3-10 Port fuel injection. (Courtesy of General Motors Product Service Training.)

calculation. The computer has already programmed into it volumetric efficiency at various RPM's, and the displacement of the engine. It also finds out through sensors manifold air temperature and pressure. Knowing these variables and performing the calculations, the computer is able to derive mass air flow.

The other method simply uses a mass air flow sensor, or a temperature compensated vane air flow meter (their operation is discussed in Chapter 2). A signal is sent by the mass air flow sen-

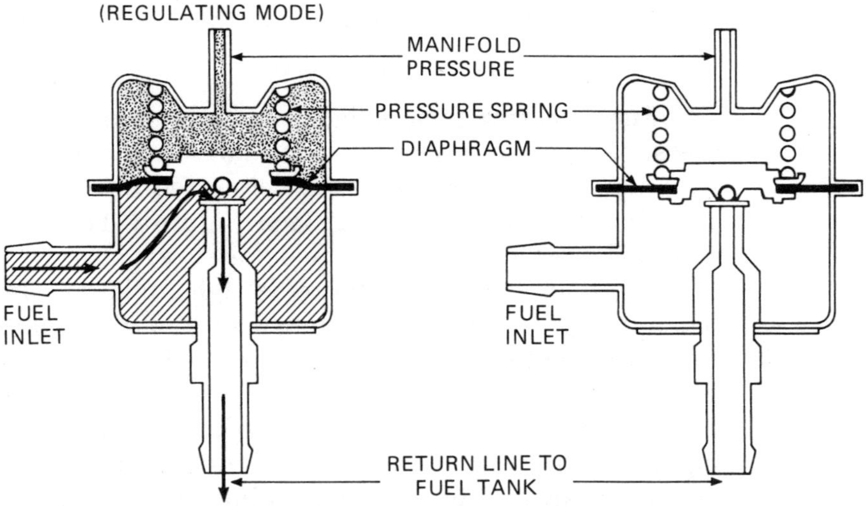

Figure 3-11 Vacuum controlled fuel pressure regulator. (Courtesy of General Motors Product Service Training)

sor to the computer which converts the signal into the mass air flow rate.

Obviously, the more air entering the engine the longer the fuel injectors will stay on. The computer not only has to know the mass air flow rate, but also throttle position, engine temperature, and exhaust oxygen content in order for it to determine how long the fuel injectors are on. Not all injectors are turned on at once (except for starting). On nearly all engines half the injectors will be energized per revolution, the other half the next revolution.

During starting, all fuel injectors are turned on to provide a rich mixture. Some injection systems use a cold start injector (valve) which is activated by a temperature sensor. If it is cold enough, the cold start valve is activated, injecting more fuel into the air intake.

Port fuel injection also eliminates a need for heated air intake. Because the fuel is injected just before the intake valve, it does not form droplets in the intake tract. When the fuel forms droplets, emission and cold weather performance suffer. Engine coolant is circulated around the throttle valve on most systems to prevent the throttle valve from freezing.

3-7-3 Sequential Fuel Injection

Sequential fuel injection is nearly identical to port fuel injection but with one important difference. The fuel injectors are turned on according to the engine firing sequence. The computer is able to determine number one cylinder position by the different pulse width sent by the Hall effect switch. The computer now controls each individual injector, which means the on-time for each injector can be different. This finer control over the air/fuel mixture lowers emissions and improves fuel economy and performance.

§ 3-8 ENGINE COMPUTER TESTING

This section will cover the guidelines and basic test procedures used in diagnosing engine computer systems. The test procedures are meant to guide you in and explain to you the methods of testing. *Always consult the proper service manual for specifications and diagnostic routines.*

Before taking on a customer complaint, be sure that you

- know the system
- have the proper tools
- know the manufacturer's diagnostic procedure and specifications

When approaching a complaint regarding a computer-controlled engine, it is important to realize that the computer system may not be at fault. Get the facts from the owner. Have the owner describe the problem and under what conditions the problem occurs. Next, road test the car to verify the complaint. Try to duplicate the conditions that the owner described. For intermittent problems, the car may have to be driven for an extended period or left overnight. Explain this to the owner to prevent misunderstandings later.

Once the complaint is verified, diagnosis of the problem can begin. *Do not forget the basics.* For example, if the complaint is poor performance, it could be the result of the basic items needing attention such as spark plugs, a vacuum leak, faulty thermostat, restricted exhaust, dirty fuel injectors, poor quality fuel, or low compression. In other words, the engine, the fuel system, and the ignition system must be in good condition. Attempting to diagnose the computer system before taking care of any of these systems will consume a lot of time and cause much aggravation. Check the engine, fuel, and ignition systems. Visually check the condition and proper routing of the vacuum hoses.

When making a visual inspection of the computer system before testing, *never disconnect connectors or wiggle wires.* Actions such as these might temporarily eliminate the problem and make subsequent diagnosis more difficult. Always run computer system tests before disturbing anything.

Many manufacturers' computer systems provide some form of a system performance check. This is a method where the computer system is put into a mode where it checks itself out. You will need a manufacturers' service manual or one of the computerized engine control manuals since every manufacturer has its own unique way of diagnosing its computer system. Should trouble occur, it may be represented as a trouble code. A trouble code is a number stored by the computer that indicates to the technician a circuit may be at fault or something is causing the circuit to set the trouble code. Disconnecting a sensor will activate a trouble code. Trouble code numbers are set when the computer receives a value out of the normal range for a particular circuit. It is important to note that not all trouble codes set instantly. They may require a few minutes or a certain mode of operation before they set. For example, you have put the computer into its diagnostic routine according to the service manual. A trouble code number 20 has appeared. According to the service manual, there may be a fault in the engine coolant temperature sensor circuit. It does not indicate that the sensor itself is bad, just that there may be a fault in the circuit or something causing the circuit to give a value out of its normal range. You will have to check it out further. You will probably be told first to check the coolant level and thermostat operation. If this is satisfactory, you will troubleshoot further with a high impedance (resistance) digital volt-ohm meter.

A digital volt-ohm meter is needed in computer circuits because of its high impedance (see Figure 3-12). When checking for continuity with a voltmeter, an extra branch is created in the circuit. And following parallel circuit characteristics, this extra branch lowers resistance and increases current flow. A low impedance voltmeter may not offer enough resistance, which may cause it to give inaccurate readings, leading to false conclusions, or it may damage the computer circuit. With a high impedance digital volt-ohm meter, the internal resistance of the meter is so high that its effect on the circuit is virtually nonexistent. A test light is not recommended for tests on computer circuits.

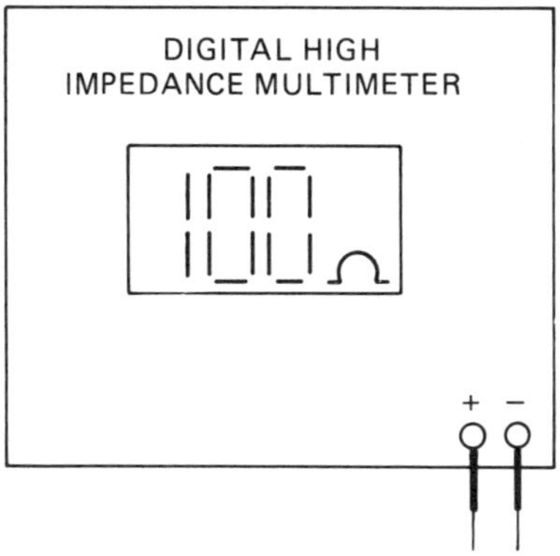

Figure 3-12 Digital volt-ohm meter.

3-8-1 Basic Sensor Tests

Basic tests of sensors are easy if you know how they work. The following tests describe the testing procedures. Consult a service manual for specifications.

3-8-1-1 Thermistor Sensor Tests. A thermistor sensor and circuit are checked in the following manner.

First, the thermistor circuit is checked for the proper reference voltage and continuity. To do this, disconnect the sensor and connect a digital voltmeter to the leads in the connector from the computer (see Figure 3-13a). The reading should be approximately 5 volts (on some early systems this could be 9 volts).

Resistance of the thermistor is measured by connecting a digital ohm meter to the thermistor sensor. (Note: anytime the resistance of a device is measured it must be isolated from the rest of the circuit) (see Figure 3-13b). Record the resistance. The reading in ohms should correspond to the temperature of the thermistor (see Figure 3-13c).

A voltage drop will provide a reading of the voltage signal as measured by the computer. Connect a set of jumper wires from the sensor to the connector, or use straight pins. Connect the voltmeter in parallel to perform a voltage drop test. Start engine. The voltage drop, which is the voltage signal the computer reads, should change as temperature changes (see Figure 3-14).

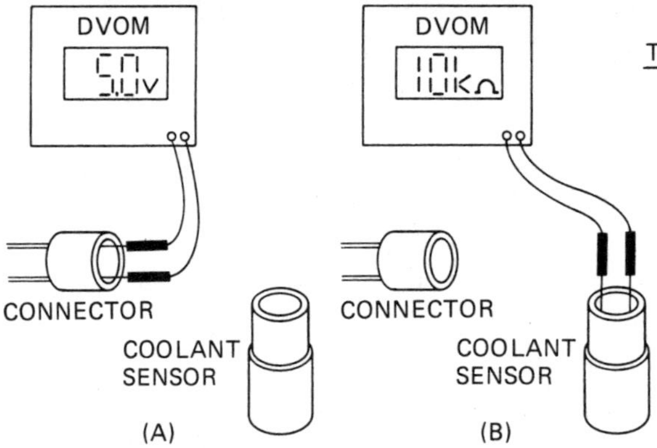

COOLANT SENSOR

TEMPERATURE	RESISTANCE
250° F	1 kΩ
150° F	10 kΩ
100° F	30 kΩ
50° F	90 kΩ
0° F	150 kΩ
-50° F	250 kΩ

(C)

Figure 3-13 (a) Checking reference voltage. (b) Checking sensor resistance. (c) This chart is provided as an example of how change in temperature affects resistance. *Do not use this chart for specifications.* Consult the service manual.

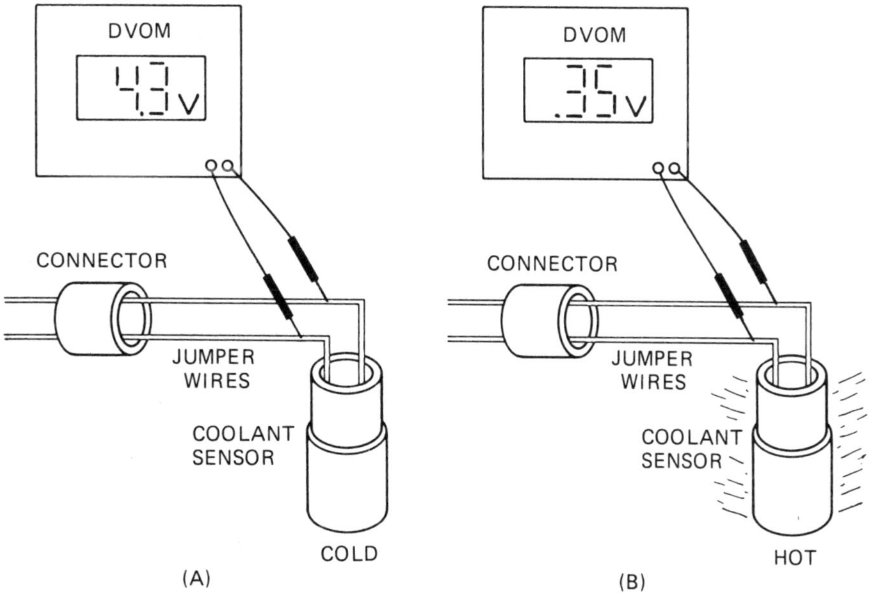

Figure 3-14 Checking voltage drop. (a) When sensor is cold. (b) When sensor is hot.

The following is an example of checking the engine coolant temperature sensor (which is a thermistor). In Figure 3-15 the engine coolant temperature (ECT) sensor is connected to the computer at pins 46 and 7. You can locate the ECT on the car by referring to the wire color chart in the service manual and tracing the wires in the wiring harness. To measure reference voltage, disconnect the connector at the ECT and connect the digital volt-ohm meter to the wires leading to the computer (Pins 46 and 7). Read the reference voltage and compare to specifications.

Next, connect the digital ohmmeter to the ECT to measure resistance. Note the reading. Also, measure engine temperature. The resistance and temperature should match specifications.

To measure voltage drop across the ECT, connect the digital voltmeter in parallel to wires 46 and 7. Start the engine and record the voltage. Measure engine temperature and compare both readings to manufacturer's specifications.

3-8-1-2 Potentiometer Sensor Tests. There are three leads to a potentiometer. Connect a voltmeter from the voltage reference line to the ground line. Reading should be 5 volts (nine volts on some systems) (see Figure 3-16).

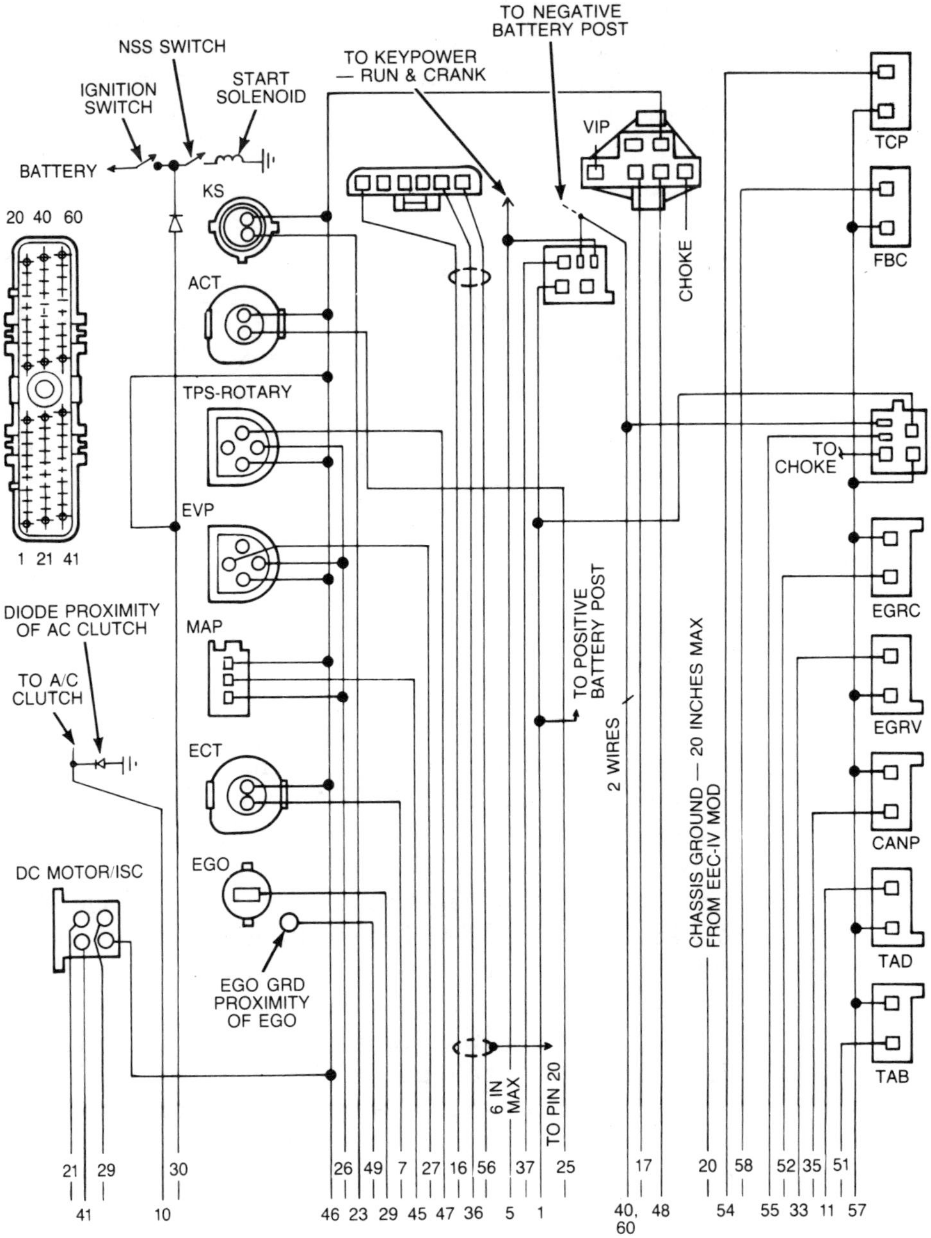

Figure 3-15 (Courtesy of Ford Motor Company)

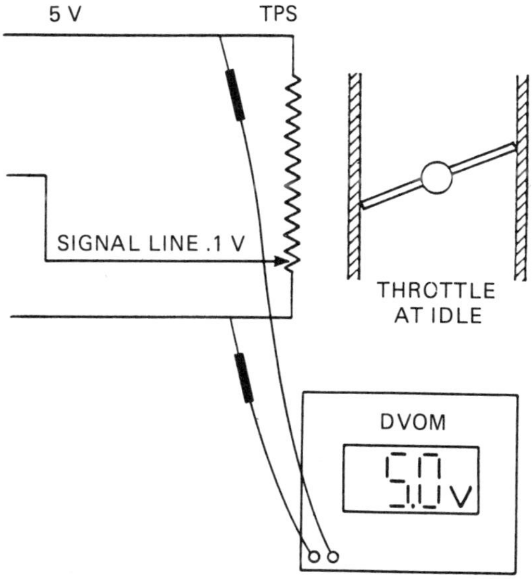

Figure 3-16 Potentiometer sensor test.

Next, with the connector removed, *on the sensor connector* connect an ohmmeter to the signal line and to either the ground or voltage reference line as indicated by the service manual. Ohmmeter readings should change as the signal arm on the potentiometer is moved.

A voltmeter can be used to measure the voltage signal which is the voltage drop across the resistor. With the circuit connected connect a voltmeter to the signal wire and ground line. Move the signal arm and note the readings as the signal arm changes position. Readings should correspond with the movement of the signal arm (see Figure 3-17).

A throttle position sensor (TPS) is used in the following example to demonstrate the testing of a potentiometer. The TPS shown in Figure 3-18 is connected to Pin 26 (VREF, reference voltage), Pin 46 (signal return-ground), and PIN 47 (throttle position signal). The proper wire colors could be identified by using charts in the shop manuals. A digital voltmeter would be connected between the voltage reference line (Pin 26) and the ground line (Pin 46) to measure reference voltage. The next step, measuring total resistance, a digital ohmmeter is connected to the reference voltage line (Pin 26) and ground line (Pin 46) at the TPS with with the con-

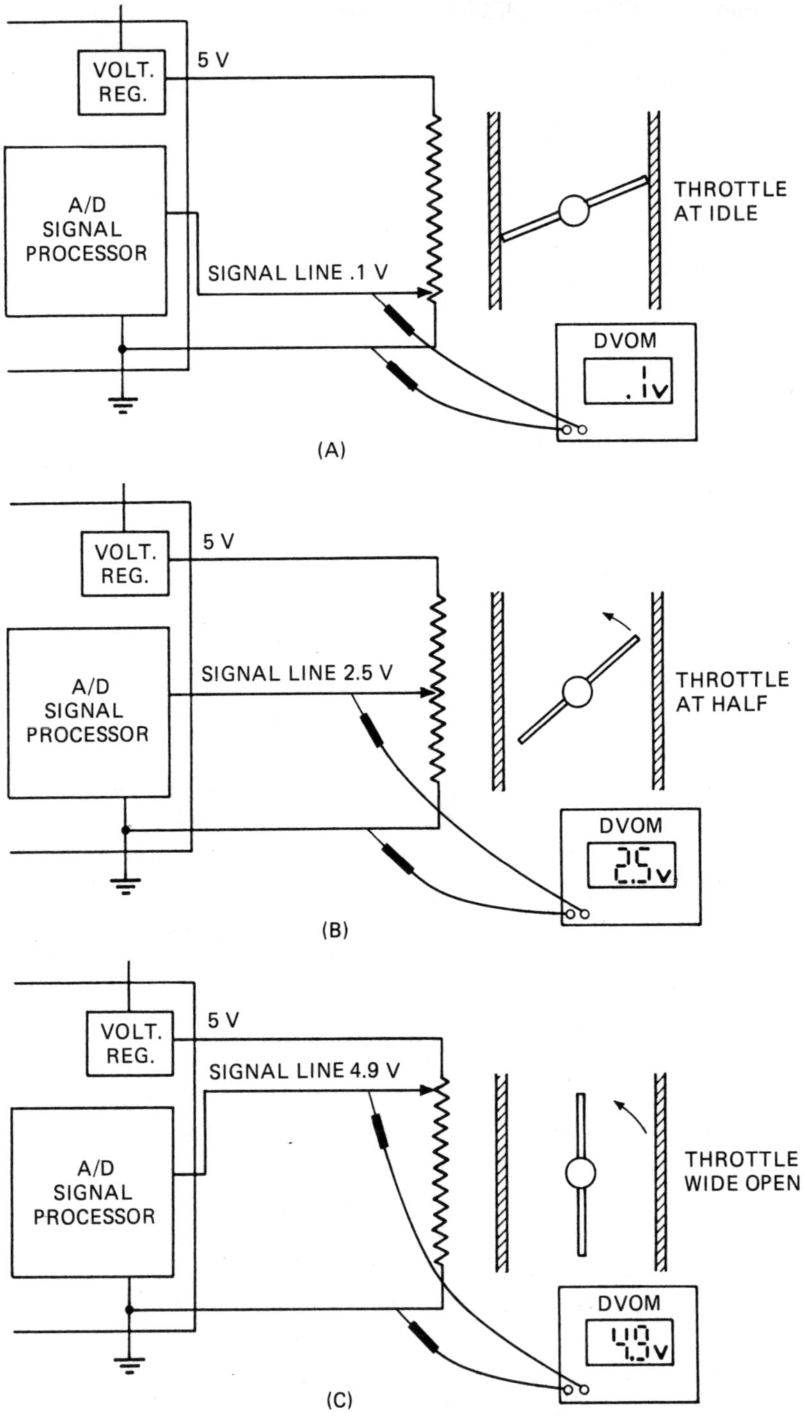

Figure 3-17 Throttle position sentor. (a) Throttle at idle. (b) Throttle at half. (c) Throttle wide open.

nector to the computer disconnected. To measure voltage drop, which is the throttle position signal, a digital voltmeter is connected from the throttle position signal line (Pin 47) and ground (Pin 46). Move the throttle shaft and note the readings. The voltage readings should change smoothly and gradually. If they do not or if the readings are not within specifications, the TPS may have to be replaced. (Note: Some TPS are adjustable. Always refer to manufacturers' specifications.

3-8-1-3 Exhaust Gas Oxygen Sensor Tests. A conventional EGO sensor is connected to the computer with a single wire. The ground circuit is through the outer shell of the EGO to the exhaust manifold. The ground circuit is often completed by a separate ground wire attached to the engine block. Figure 3-15 is a typical wiring schematic. Pin 29 connects the EGO to the computer (processor) and pin 49 is the EGO ground.

To test the EGO, the engine must be hot and run at fast idle for two minutes prior to any testing. Then, shut the engine off, disconnect the wire from the EGO, and connect a *digital* voltmeter between the EGO terminal and an engine ground. Restart the engine and run at fast idle. The EGO should read above .6 volt. Remove a vacuum line from the engine to lean the air/fuel mixture and the EGO should go below .4 volt. If the EGO does not fall within specifications it should be inspected; it may require replacement. To monitor EGO output, with the computer circuit connected push a straight pin through wire insulation and connect a digital voltmeter between the pin and ground.

Some EGO sensors are equipped with a heating element (Chapter 2, Figure 2-14). A wiring schematic showing a heated EGO (HEGO) is shown in Figure 3-18. The same test procedure as described for the conventional (single terminal) EGO can be used but the correct terminal and wire must be selected. The individual wires can be identified by a wiring schematic showing wire colors or with a cross-reference chart (Pin numbers to wire colors) found in service manuals. Figure 3-18 shows the EGO wire going to Pin 29. The voltage test of the HEGO can be performed as described with the harness disconnected providing the engine is *hot* and run at fast idle. Remember, a lean mixture produces a low voltage signal (less than .4v); a rich mixture produces a high voltage signal (more than .6v).

2.3L HSC CFI

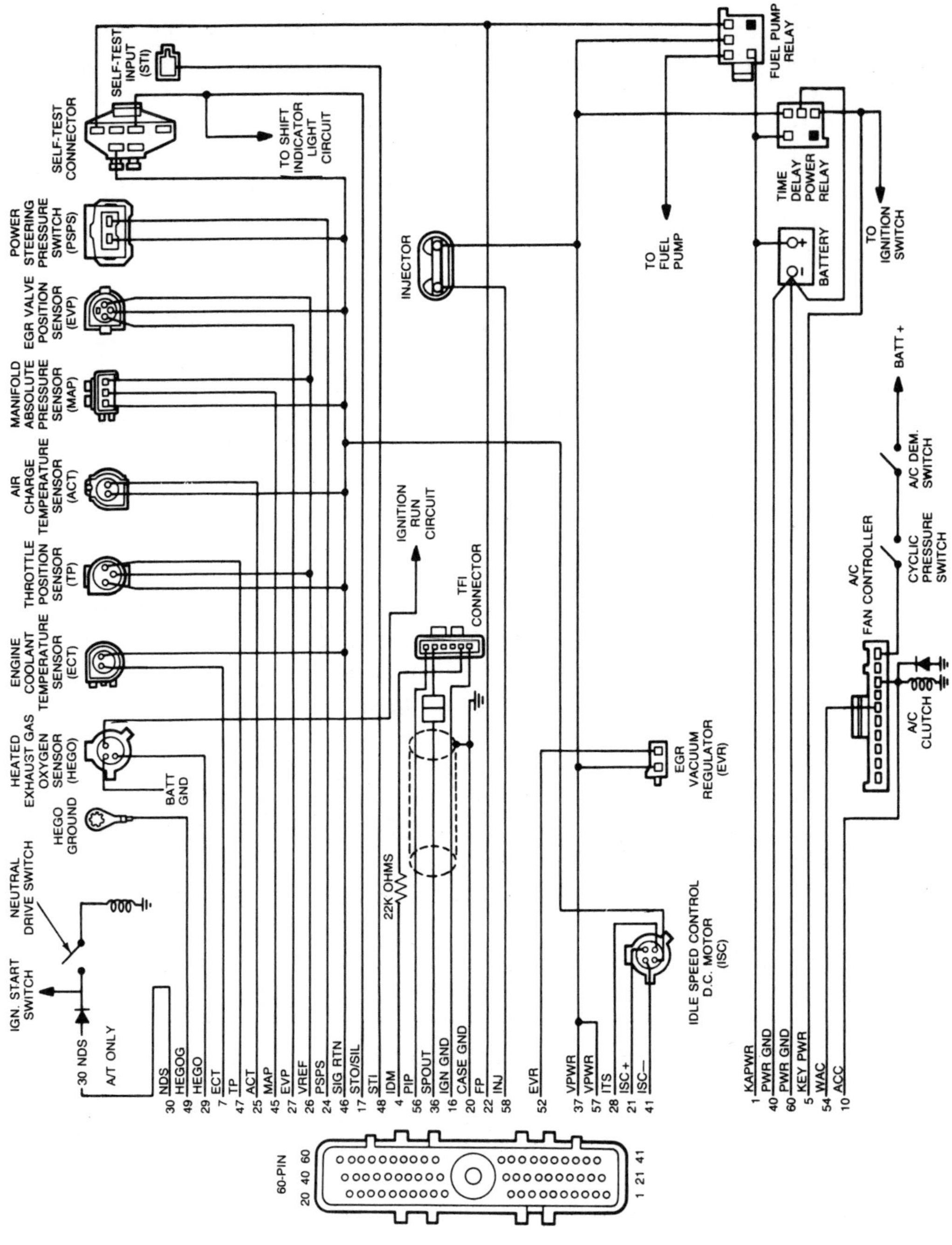

NOTE: WIRING SCHEMATIC SHOWS PIN OUT LOOKING INTO HARNESS CONNECTORS

Figure 3-18 (Courtesy of Ford Motor Company)

Caution: Never use an ohmmeter to measure the oxygen sensor. The ohmmeter power supply could damage the sensor.

3-8-1-4 MAP Sensors. MAP sensor electrical tests vary greatly between manufacturers. You will need a service manual to test the MAP sensor. However, keep in mind that the vacuum line to the sensor must be connected properly and free of leaks and restrictions. A hand-held vacuum pump is used to check for leaks. The vacuum line can become restricted by sludge accumulating within the line. Remove the line and blow through with compressed air.

3-8-1-5 Other Tests. Detailed descriptions for testing other sensors (knock, Hall effect, variable reluctance, and others) will be found in service manuals. Service manuals are indispensable for the proper test procedures and specifications. Always consult one when testing a computer-controlled vehicle.

However, we have included a series of suggestions sometimes not found in service manuals to help you in diagnosing drivability and engine performance complaints.

It is important to understand that there is no standard procedure for extracting trouble codes used in diagnosis. Even different years and models of the same car line have different requirements for tests. If procedures are not followed exactly, incorrect results and misdiagnosis may result.

Make sure the fuel pressure is to specifications. Sometimes, a road test with a pressure gauge installed is in order. The pressure should stay within specifications during all driving conditions.

Problems with fuel injectors are sometimes difficult to isolate. As a general rule, dirty injectors in high pressure systems tend to cause the engine to run lean, while dirty injectors in low pressure systems cause rich symptoms. Vehicle manufacturers and aftermarket companies have equipment and chemicals available to test and clean injectors. Good quality fuel is essential for optimum operation of fuel injected engines.

Exhaust system restrictions can cause drivability problems that are difficult to diagnose. A quick test for this is to check engine vacuum idling hot then increase RPMs to 2,500 in neutral. Engine vacuum should stay about the same as at idle. A drop in vacuum could indicate an exhaust restriction.

In a fuel injected engine equipped with a mass air flow (MAF) meter (Chapter 2, Figures 2-33 and 2-34) or vane air flow (VAF) meter (Chapter 2, Figure 2-8), no air leaks are permitted between the sensor and throttle body (see Figure 3-19), air cleaner outlet tube. Any air that enters after the air flow sensor will not be read by the computer and a lean mixture will result.

Do not overlook the operation of the cooling system thermostat. If the engine is running cooler than normal, the engine coolant sensor will report this to the computer. The computer then may provide

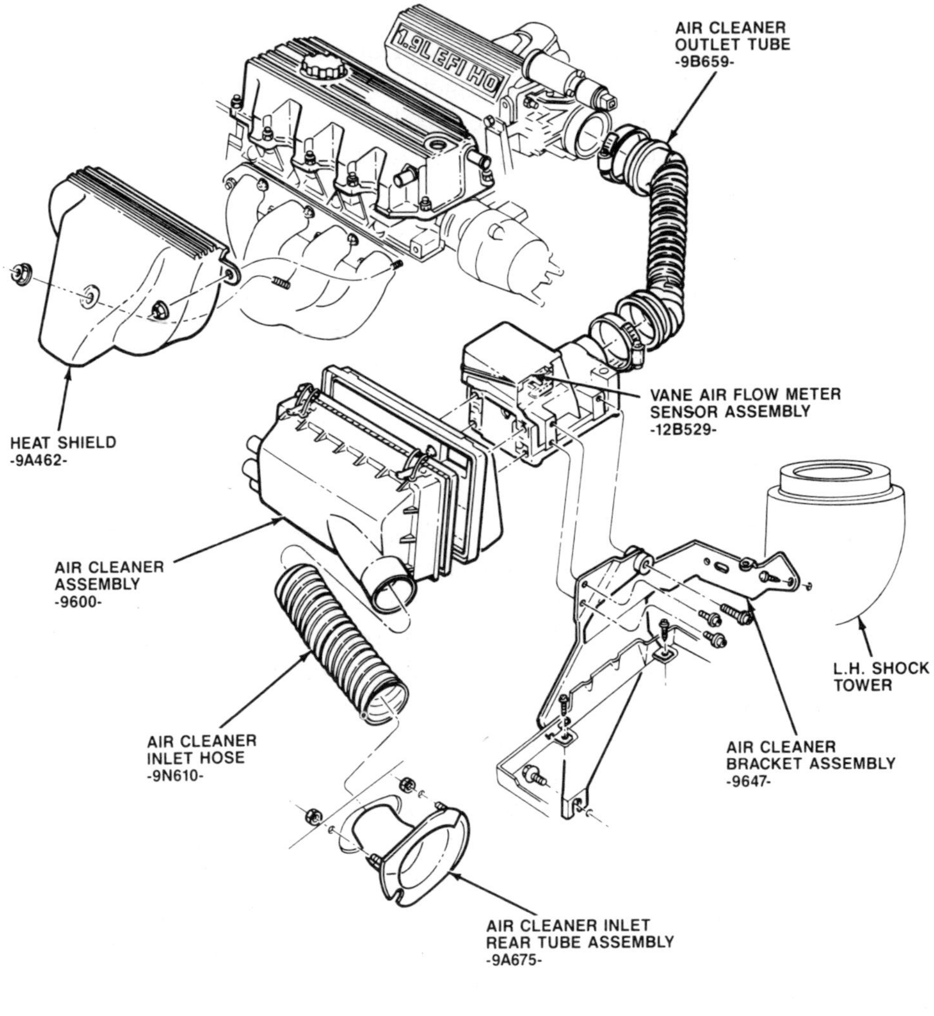

Figure 3-19 (Courtesy of Ford Motor Company)

a richer than normal mixture which could result in a reduction in mileage.

If a computer system has a heated EGO, make sure the heater is operating. In the system in Figure 3-18, there should be battery voltage between the battery end terminal and the ignition run circuit at the HEGO connector with the ignition switch on. There should also be continuity between the corresponding terminals on the HEGO sensor. Some resistance should be noted between these terminals.

Finally, the basic components must all be in satisfactory condition. Uneven compression, vacuum leaks, ignition problems, poor quality fuel, and fuel system problems may not show up on computer system tests. The computer system tests are only designed to test the control system, so the basic functions still have to be tested by conventional means.

QUESTIONS

3-1. Describe closed loop operation.

3-2. Describe open loop operation.

3-3. Describe how electronic spark advance works. What is total advance?

3-4. Where is extra air pumped to on a cold engine? A warm engine? Why?

3-5. What control devices are used on feedback carburetors?

3-6. What are the differences between throttle body and port fuel injection?

3-7. What is the main difference between port and sequential fuel injection?

3-8. Why is it important on a poor performance complaint to check the entire engine before checking the computer system?

3-9. Why is a digital volt-ohm meter needed?

3-10. What three basic tests are performed on thermistor circuits?

3-11. What meter is used to check the change in resistance of the signal arm of the potentiometer?

Chapter 4

Other Automotive Computer Systems

OBJECTIVES

After studying this chapter, the student will be able to

- describe what electronic instrumentation must accomplish
- name the two different types of speedometer sensors and describe how the circuits work
- describe how the temperature sensing circuits work
- name the two different types of oil pressure sensors and describe how the circuits work
- describe how the fuel level circuit works
- describe how the computer monitors system voltage
- describe the two different types of tachometer sensors and describe how the circuit works
- describe a simple series circuit warning indicator
- describe two different types of electronic warning indicators
- describe how a series circuit information center works
- describe how a Wheatstone bridge information center works
- describe how the LED, VTF, and LCD operate
- describe anti-lock braking operation
- identify the anti-lock brake system inputs and outputs and describe their functions
- describe air suspension operation
- identify air suspension inputs and outputs and describe their functions
- describe speed control operation

§ 4-1 ELECTRONIC INSTRUMENTATION

For several years mechanical gauges, electrical gauges and warning lights were the only means available to relay information to the driver. Now, electronic or digital dash instruments provide another way to communicate information to the driver about the environment and engine operation. They must be dependable, easy to read, and accurate. There are a variety and combination of different displays, but essentially electronic instruments are composed of three major sections; inputs, the computer, and outputs. Just like engine computer operation, the electronic instrument computer uses input information from sensors to make decisions. Unlike the engine computer, the dash computer will turn on the proper display, alphanumeric characters, or a voice telling the driver what is happening. The computer samples each circuit sequentially. The following section will explain how each electronic instrumentation operates. It is important that Chapters 1 and 2 are thoroughly understood, particularly series circuit characteristics and sensor operation.

4-1-1 Electronic Speedometers

There are two types of input circuits used with electronic speedometers; the permanent magnet (PM) buffered circuit and the photoelectric (optical) buffered circuit.

The PM buffer circuit consists of a PM generator, a buffer, fixed resistors wired in series, and a computer. As discussed in Chapter 2, the PM generator produces an AC voltage. The faster the PM generator spins, the higher the frequency (see Figure 4-1). The PM generator is located in the transmission and is geared to the output shaft of the transmission (see Figure 4-2). It is wired to the buffer. As the polarity of the AC voltage signal reaches maximum, the transistor in the buffer is turned on, producing a low voltage signal after R1 (high voltage drop) (see Figure 4-3). As the AC signal changes polarity and approaches its minimum, the transistor in the buffer is turned off, which produces a high voltage signal across R1 (little voltage drop). The computer can then count the number of high/low voltage pulses over a given period and interpret this as vehicle speed (see Figure 4-3b, 4-3c).

The photoelectric (optical) circuit consists of an LED, photocell, and control circuitry. The LED emits an infrared light beam which

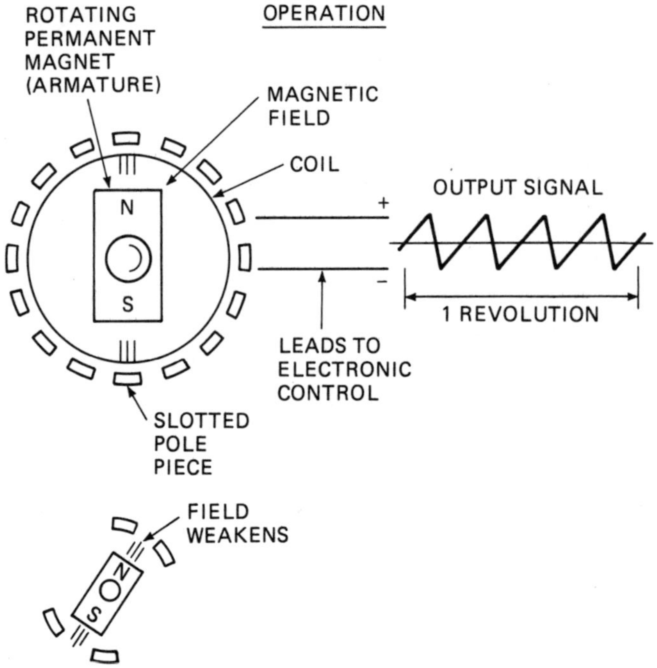

Figure 4-1 Electronic speedometer. Permanent magnet generator.

is invisible to the eye. If this light beam were to reach the photocell, the photocell would turn on. Two basic setups are used. One method is to have the photocell opposite the LED with a slotted rotor, or wheel, in between (see Figure 4-4). The rotor is attached to the speedometer cable. As the cable rotates the rotor, a slot within the rotor will allow the light beam to pass, turning on the photocell. Further rotation will move the slot out of the beam of light and the light

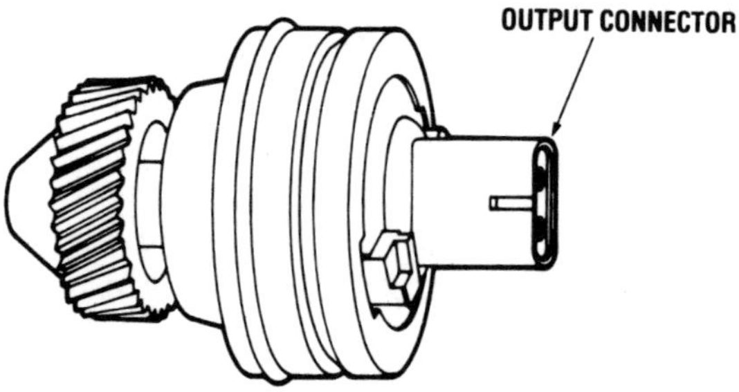

Figure 4-2 Permanent magnet generator vehicle speed sensor (Courtesy of General Motors Product Service Training).

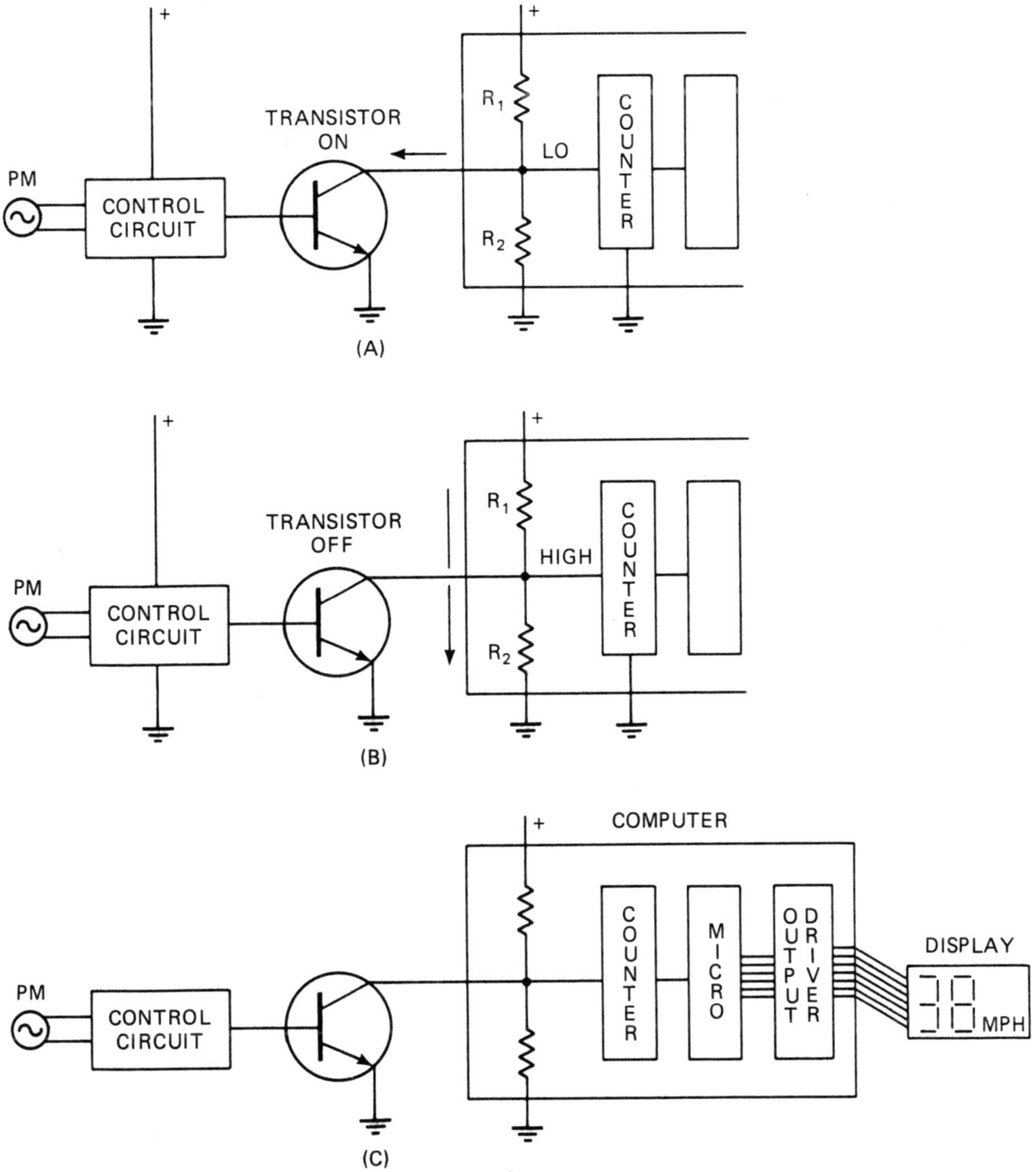

Figure 4-3 Electronic speedometer. Permanent magnet buffered circuit. (a) Transistor turned on. (b) Transistor turned off. (c) From the counter the microprocessor can determine vehicle speed and turn on the proper output driver, which will light the display.

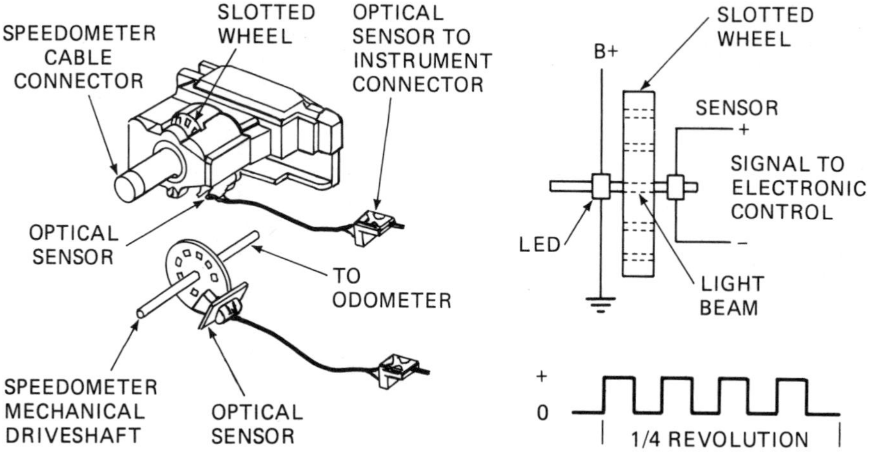

Figure 4-4 Photoelectric circuit (Courtesy of Ford Motor Company).

beam will no longer reach the photocell. The other basic setup uses reflective blades mounted on the speedometer cable (see Figure 4-5a). As the cable rotates, the reflective portion of the blade will reflect light back to the photocell, turning on the photocell. Further rotation moves the reflective blade out of position, the light does not reach the photocell, and the photocell turns off. The control circuitry is similar to the buffer circuitry (see Figure 4-5b), the major difference being that the transistor is turned on by the weak digital voltage signal of the photocell. The computer counts the number of high/low voltage pulses caused by the change in voltage drop across the sensing resistor (see Figure 4-6). Again, the number of pulses in a given period determines vehicle speed.

The speedometer readout can be given out with the conventional needle movement or by digital display. Needle movement, though it is directly controlled by the computer, is conventional in appearance (see Figure 4-7). Digital displays provide a lighted number to the driver (see Figure 4-8). How the lights themselves work is discussed later in this chapter.

The number of pulses is also used to calculate distance. In the conventional odometer, gears turned by the speedometer cable caused the numbered wheels to turn, indicating miles traveled. With electronic speedometers, the odometer can be mechanical or digital. The mechanical type uses a stepper motor to turn the

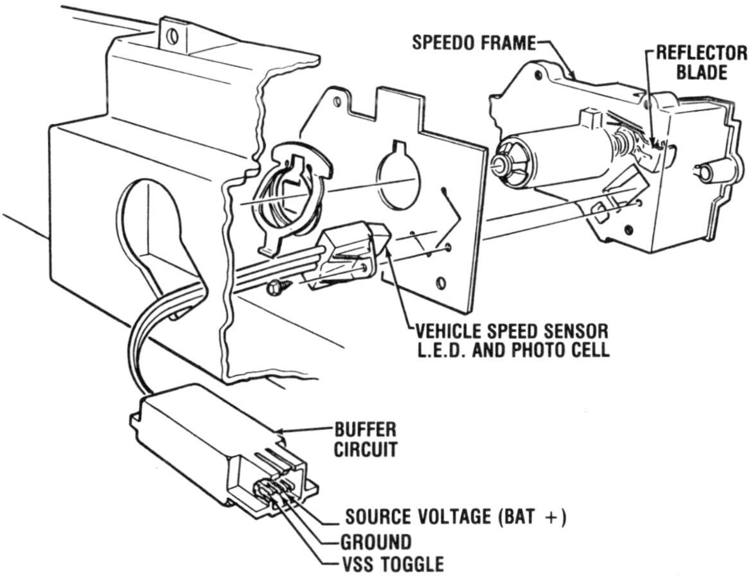

(a)

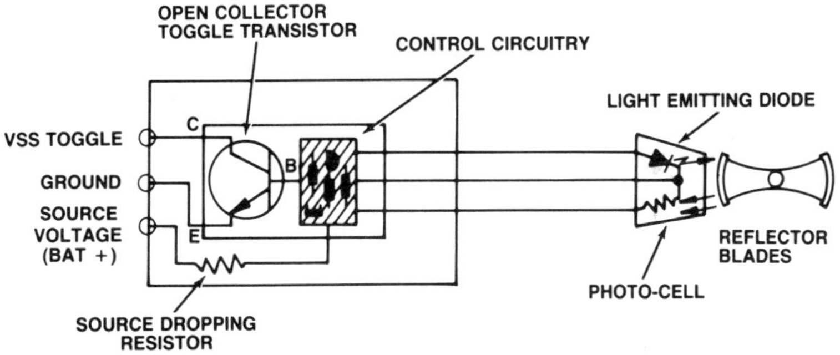

(b)

Figure 4-5 (a) Photoelectric vehicle speed sensor. (b) Optical buffer circuitry (Courtesy of General Motors Product Service Training).

mechanical number wheels. The stepper motor is pulsed on/off, which rotates the wheels (see Figure 4-9). The number of pulses is directly proportional to vehicle speed. The digital odometer uses the computer to turn on the correctly numbered display, indicating miles traveled. A special chip, called the nonvolatile random access

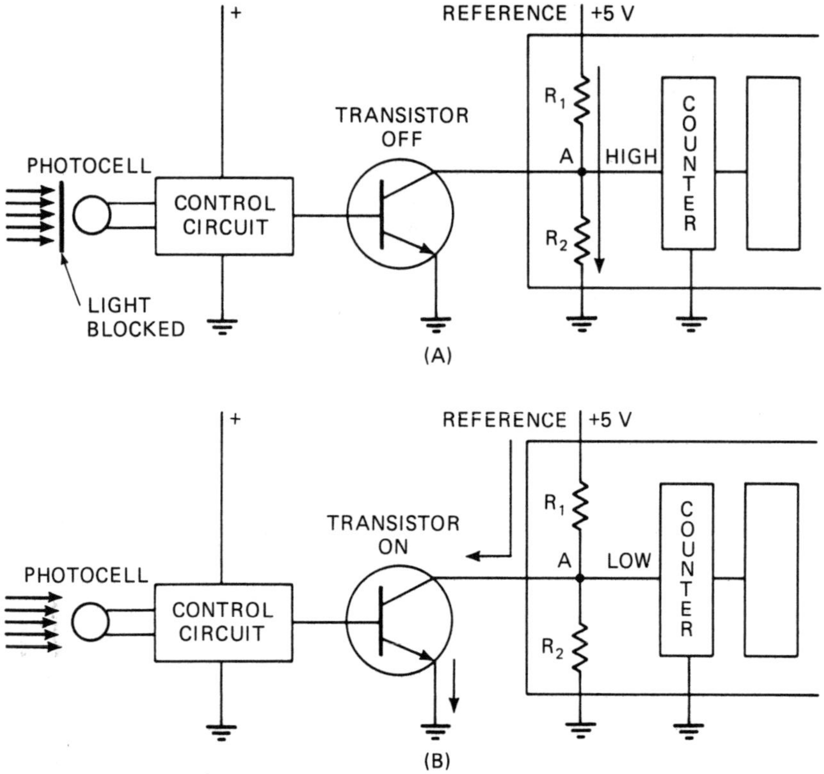

Figure 4-6 Photoelectric circuit. (a) With the light blocked from the photocell, the transistor is turned off. Current flows across R1 and R2; voltage at point A would be 2.5 volts (a high signal). Note: R1 and R2 have equal resistance. (b) With the photocell exposed to light, the transistor is turned on. This will connect point A to ground (R2 is bypassed) and will result in zero volts (a low signal) at A.

memory (NVRAM) chip, which can retain its memory when power to the computer is turned off, is used to store the number of miles traveled. Should this chip be damaged, the replacement NVRAM will indicate zero miles traveled. Generally, a warning light on the dash will indicate that this chip has been replaced, and a sticker located on the car should record the number of miles traveled on the car before the replacement.

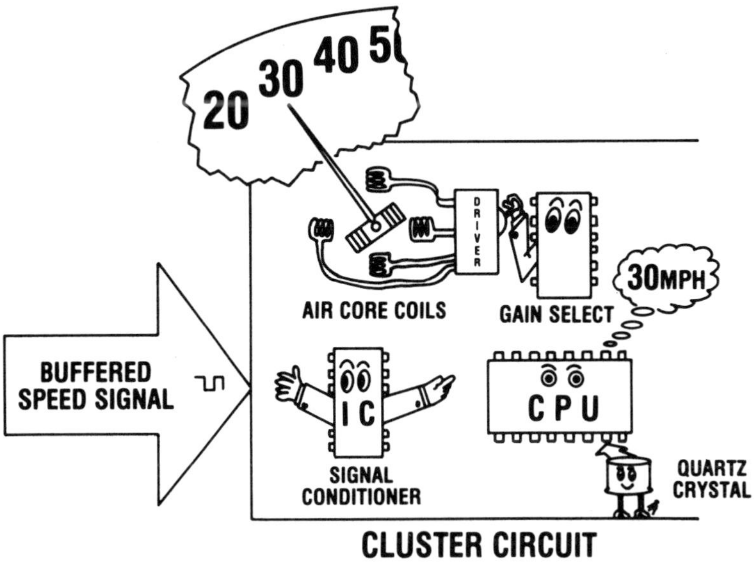

Figure 4-7 Quartz-electric cluster (Courtesy of General Motors Product Service Training).

4-1-2 Temperature Sensor Circuits

Temperature sensing circuits are used to indicate engine coolant, oil, and air temperatures. Thermistors are used in the appropriate location to sense temperature. Operation of these circuits is identi-

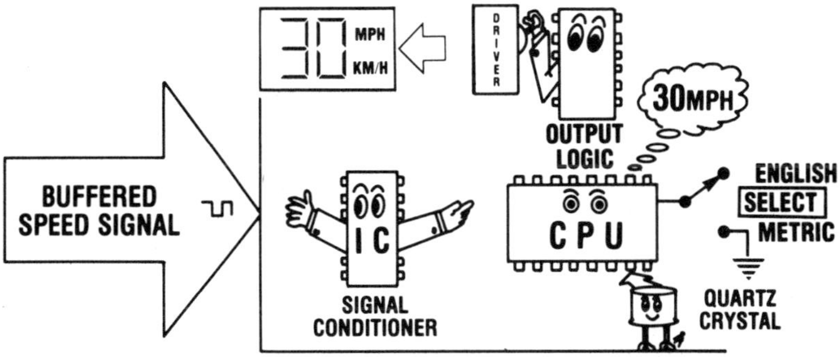

Figure 4-8 Digital cluster circuit (Courtesy of General Motors Product Service Training).

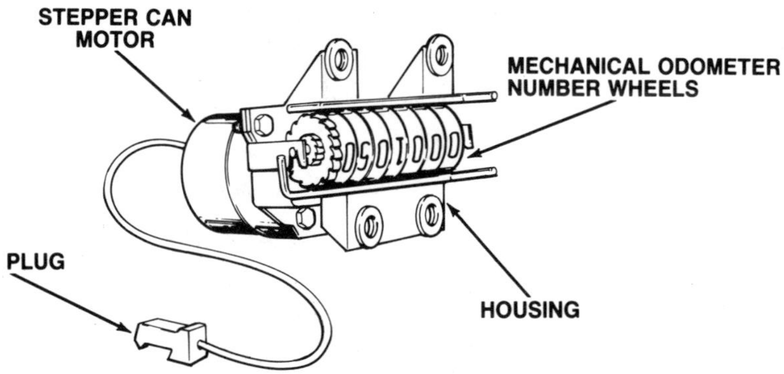

Figure 4-9 Stepper motor odometer (Courtesy of General Motors Product Service Training.

cal to those explained in Chapter 2. A thermistor is connected in series to a fixed resistor with an A/D converter connected between the fixed resistor and the thermistor. As the thermistor warms up, its resistance decreases, the A/D converter measures the decreasing voltage drop and sends the appropriate signals to the microprocessor (see Figure 4-10). The microprocessor will turn on the appropriate number that represents the temperature. In some cases a warning light may come on if normal limits are exceeded.

4-1-3 Oil Pressure Sensor Circuits

There are two types of sensors used for input to the computer. The common and familiar type of sending unit uses a rheostat, the other type a piezoresistor. Both are variable resistors. The rheostat uses a wiper arm attached to diaphragm. As oil pressure increases, the wiper arm moves further up the resistor, increasing electrical resistance. In a piezoresistor sending unit, the piezoresistor is between a diaphragm and the sending unit housing. As oil pressure flexes the diaphragm, pressure on the piezoresistor increases, which also increases electrical resistance.

Except for the difference in sending units, the circuitry and electrical operation is the same. The sending unit is connected in series to a fixed resistor; an A/D converter is connected between the fixed resistor and the sending unit. As oil pressure increases, the

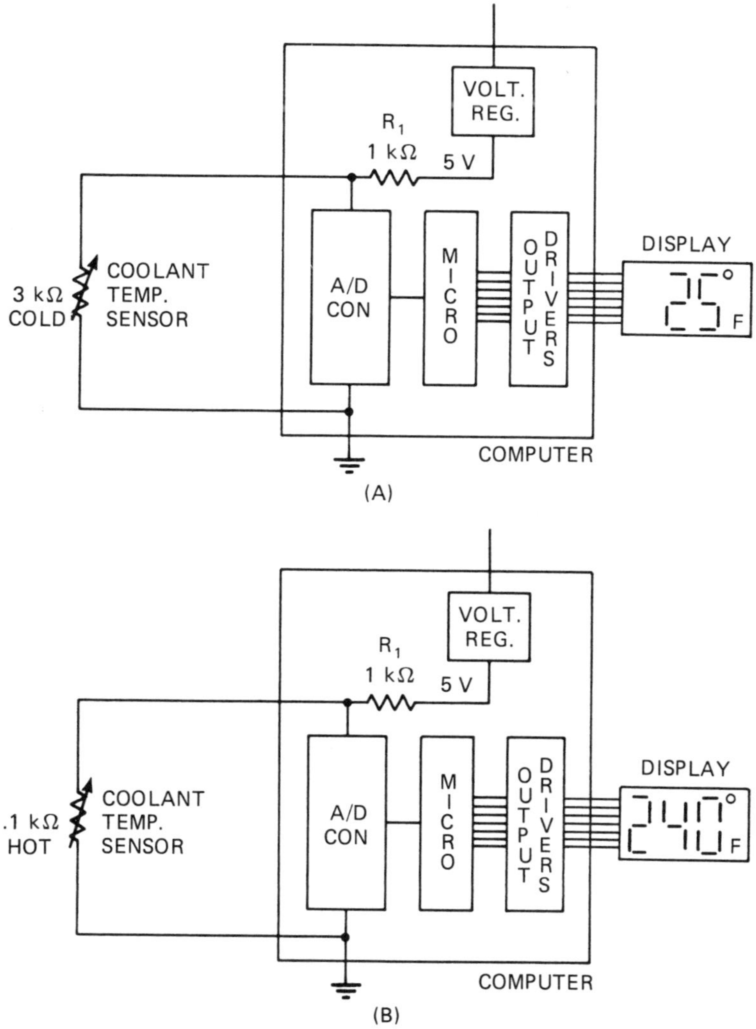

Figure 4-10 Temperature sensing circuit. (a) With the coolant temperature sensor cold and its resistance high, the A/D converter will read a high voltage drop across the sensor and pass this information on to the microprocessor. The microprocessor will turn on the appropriate drivers lighting the display. (b) When the coolant sensor is hot, both resistance and the voltage drop decrease. The A/D converter will pass the information to the microprocessor which will turn on the appropriate output drivers, lighting the display indicating temperature.

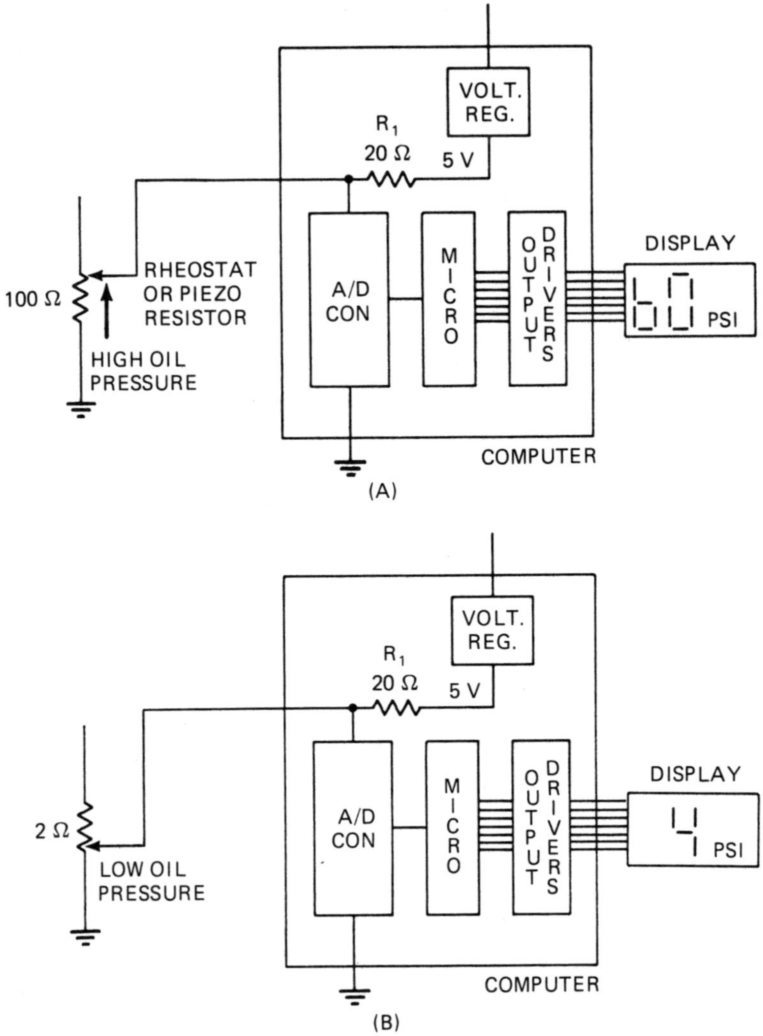

Figure 4-11 Oil pressure sensor circuit. (a) Under high pressure the oil pressure sending unit is at high resistance. A high voltage drop is read by the A/D converter and this information is passed on to the microprocessor. The microprocessor will process the information and turn on the output drivers which will light up the display indicating oil pressure to the driver. (b) A low oil pressure and therefore a low voltage drop. The computer will process the information as described previously and light the display.

sending unit increases in voltage drop as measured by the A/D converter. The microprocessor receives the information from the A/D converter and will light the appropriate numbers to indicate the engine oil pressure (see Figure 4-11a, 4-11b).

4-1-4 Fuel Level Sensor Circuits

The sending unit for digital display of the fuel level is the familiar rheostat. It can be set up two ways:

- as fuel level increases, resistance decreases
- as fuel level increases, resistance increases

Again, there is a fixed resistor in series with the sending unit and an A/D converter connected between the fixed resistor and sending unit. The A/D converter measures the change in voltage drop as the fuel level and therefore resistance changes and sends the corresponding information to the microprocessor (see Figure 4-12).

The microprocessor, however, does not immediately change the digital display that represents the fuel level. The fuel level is constantly changing because of acceleration, braking, and turning. This would cause the digital display to move rapidly, making it difficult for the driver to determine the fuel level. The computer program commands the microprocessor to ignore the rapid fuel fluctuations and respond slowly to fuel level changes. This results in a gradual change in the digital display of the fuel level.

A warning is often provided to tell the driver when the fuel level is low. When the sending unit reaches a predetermined level, the computer will turn on a light, informing the driver.

4-1-5 Electrical System Voltage

No external sensor is used to monitor system voltage. The computer is continuously reading the voltage it receives. System voltage is fed into an A/D converter which in turn sends a corresponding signal to the microprocessor (see Figure 4-13). From this point it turns on a display telling the driver the system voltage.

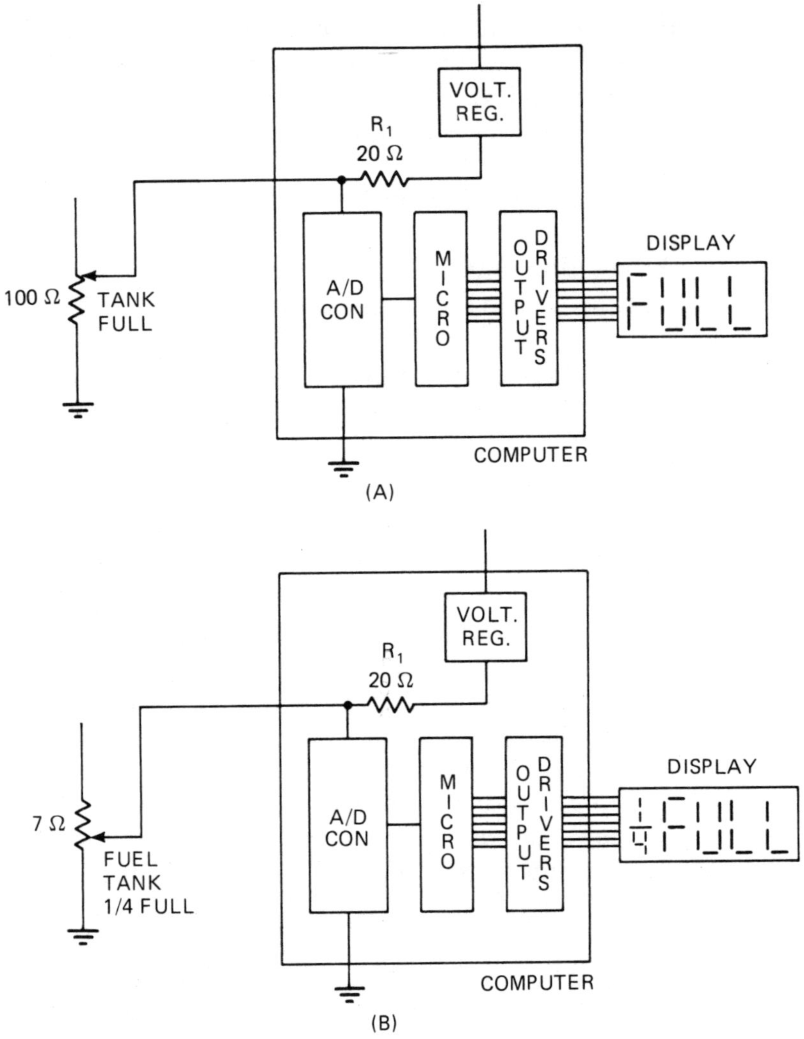

Figure 4-12 Fuel level sensor circuit. (a) Tank full. The fuel tank sending unit is designed so that a full tank of fuel has the maximum resistance. The high voltage drop reading is processed, and the display telling the driver the fuel tank is full is lit. (b) As the fuel level drops, sending unit resistance drops. The low voltage drop signal is interpreted by the computer, and the display is lit indicating the amount of fuel left in the tank.

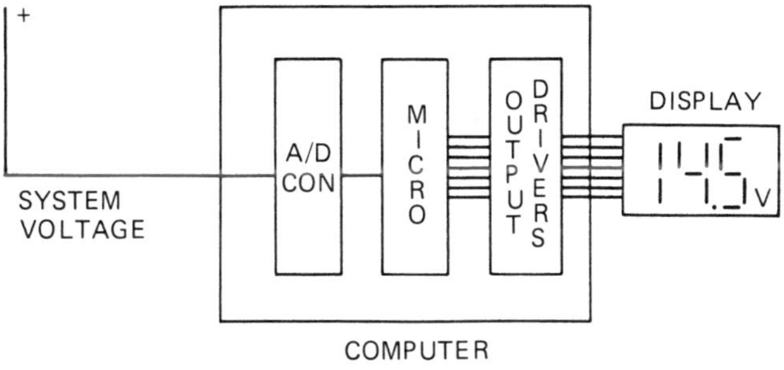

Figure 4-13 Electrical system voltage.

4-1-6 Tachometer

The computer receives the tachometer signal from either the distributor or crankshaft position sensor. The input sensor itself is either a Hall effect or variable reluctance type (the variable reluctance sensor is used with a buffered circuit to provide a digital signal). Operation of these circuits is the same as described in Chapter 2. The electronic dash computer will take the signal and light a graphic display or present a number that indicates engine RPM (see Figure 4-14).

4-1-7 Warning Indicators (Telltales)

Warning indicators (sometimes called warning lights) are designed to tell the the driver when a system will not or soon will not function properly. The indicators light when a system is no longer normal. Coolant temperature, oil pressure, low fuel level, hydraulic brake system failure, and system voltage often use an indicator that will light when when a system is out of range. Other indicators that are becoming more common are lights that tell when the

- brake, coolant, and wiper fluid is low
- door, trunk, or hood is ajar
- brake pads are worn
- headlights, parking lights, brake lights, and directional lights are not functioning

Warning indicators fall into two categories: series circuit indicators not controlled by the microprocessor, and microprocessor controlled indicators (sometimes called sequenced indicators, since

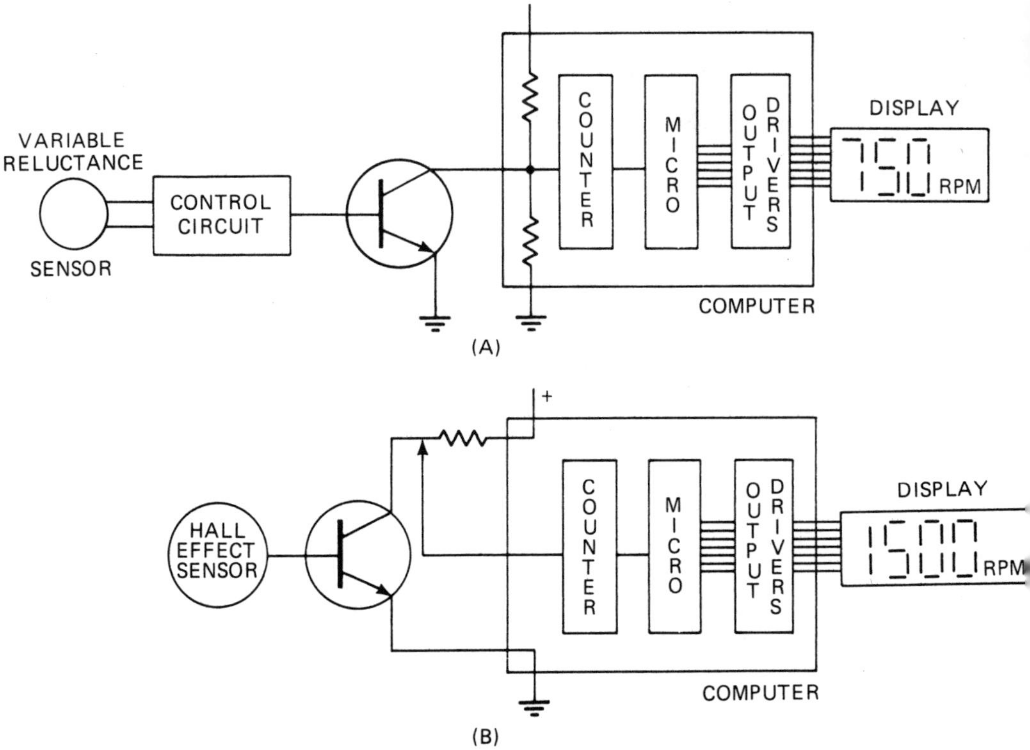

Figure 4-14 Tachometer. (a) Variable reluctance sensor. (b) Hall-effect sensor.

they will turn on only when the microprocessor determines that a system is out of range). Warning indicators can be connected both ways; it depends on the preference of the manufacturer. Circuits commonly used are described in the following section.

Measuring brake, coolant, and wiper fluid requires the use of a magnetic float with a switch in the fluid. The switch is connected in series to a warning light. When the float reaches a predetermined level, the switch will close, completing the circuit to ground and lighting the indicator (see Figure 4-15). In a microprocessor controlled circuit, the switch is connected in series to fixed resistor. An A/D converter is connected after the fixed resistor. If the fluid level is normal, the switch is open and the A/D converter sees a high voltage signal (see Figure 4-16). But when the fluid level is low and the float has reached a predetermined low level, the switch will close

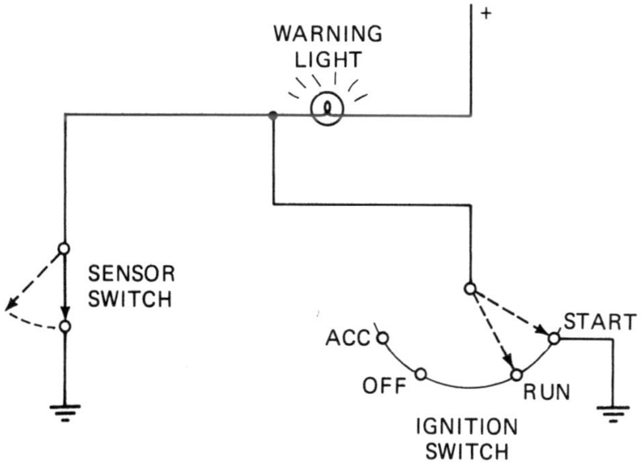

Figure 4-15 Series circuit fluid-level warning indicator.

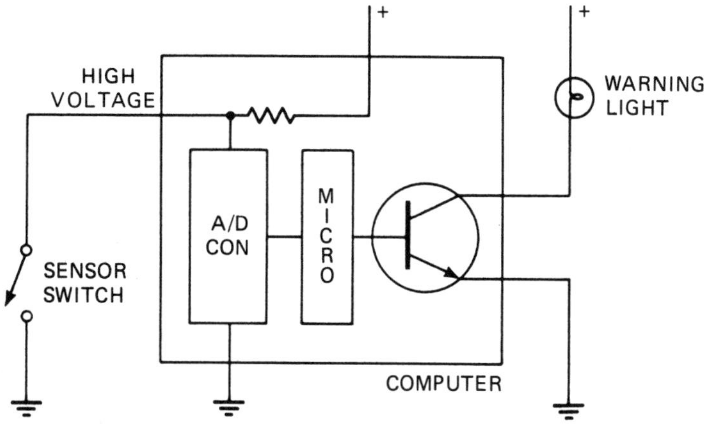

Figure 4-16 Microprocessor-controlled fluid-level warning indicator. Level is normal.

the circuit completing the path to ground. The A/D converter will read a very low voltage signal and will signal the microprocessor that the fluid level is low (see Figure 4-17). The microprocessor will turn on the proper telltale, informing the driver of the situation.

Another type of coolant sensor may use a resistive probe immersed in the coolant. When coolant surrounds the probe, its resistance is low. But when coolant is no longer present, its resistance increases, increasing the voltage drop measured by the A/D con-

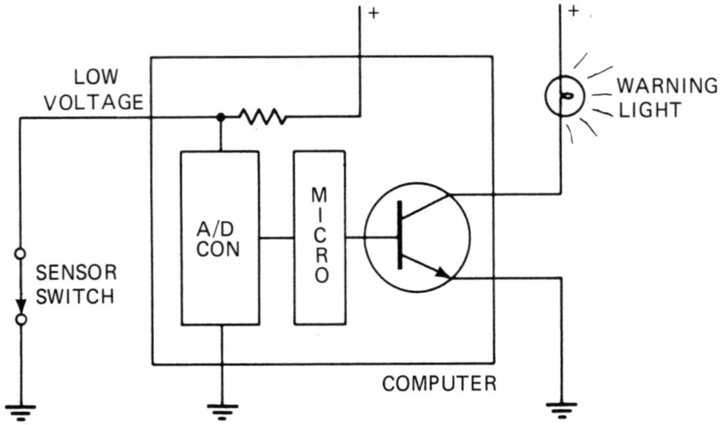

Figure 4-17 Microprocessor-controlled fluid-level warning indicator. Level is low.

verter. The A/D converter will signal the microprocessor that the coolant level is low.

The electrical circuitry for indicating if a door, trunk, or hood is ajar is also a series circuit. A warning light is connected in series with a switch. When, for example, a door is opened, the switch closes, completing the path to ground and lighting the indicator (see Figure 4-18). If these warning indicators are microprocessor controlled, a fixed resistor is connected in series to a switch. The A/D converter is connected after the fixed resistor and monitors the voltage drop across the switch. A separate circuit and switch is needed

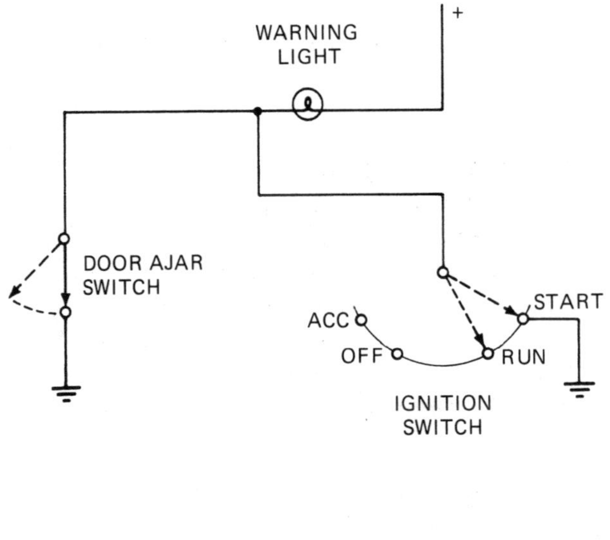

Figure 4-18 Series-circuit door-ajar warning indicator.

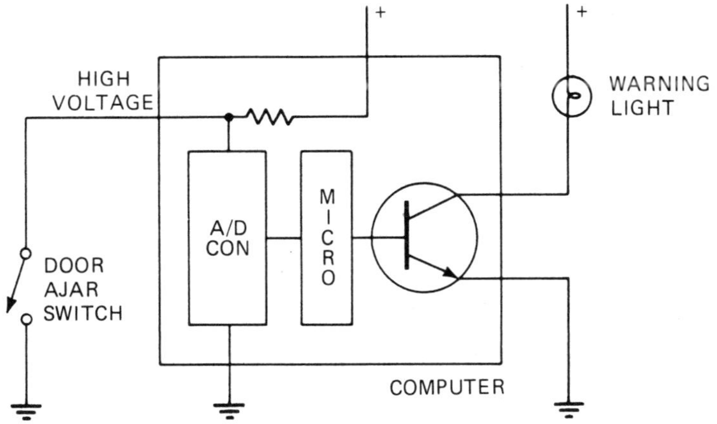

Figure 4-19 Microprocessor-controlled door-ajar warning indicator. Door closed.

on each door, trunk, and hood to determine its position. When, for example, a door is closed, the switch is open and the computer sees a high voltage signal (see Figure 4-19). When the door is open, the switch is closed, the computer sees a low voltage signal and will light the indicator (see Figure 4-20).

Brake pad wear is also sensed by a series circuit (see Figure 4-21a). In a typical example, the main components are a fixed resis-

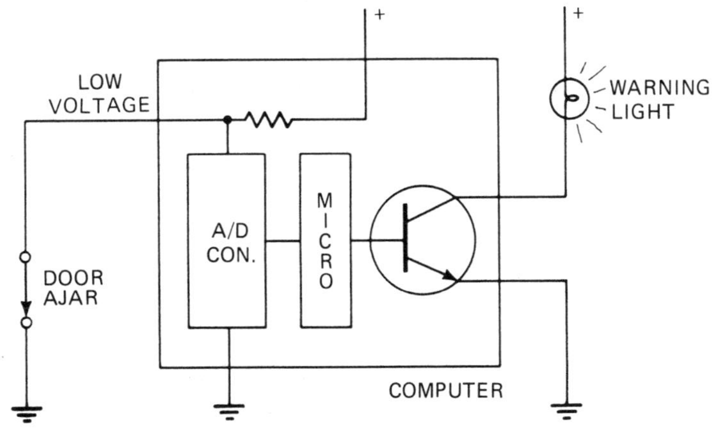

Figure 4-20 Microprocessor-controlled door-ajar warning indicator. Door open.

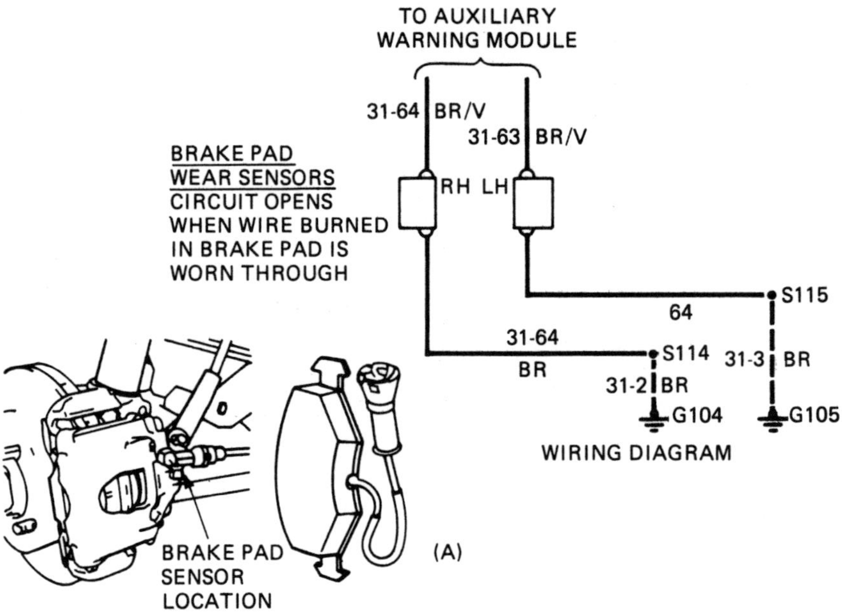

Figure 4-21 (a) Brake pad wear sensed by a series circuit. (Courtesy of Ford Motor Company.)

tor, a microprocessor, and wire in each brake pad. When normal, current flows through the fixed resistor to ground through the brake pad wire, and the microprocessor sees a low voltage reading. When the brake pad wears through the wire, the path to ground is broken and the microprocessor sees a high voltage reading. The microprocessor will turn on the brake warning light telling the driver the pads are thin and need replacing. More details are described in Figure 4-21b, 4-21c.

To help the driver know when a headlight or directional light is out, there are information centers that turn on warning light indicators. They operate on two different types of circuits.

The first type utilizes a comparitor which monitors input and output voltage across a fixed resistor located in the wiring harness (see Figure 4-22). A comparitor is able to measure and compare voltages from two different sources, then send a corresponding signal to the microprocessor (the comparitor is also called a high speed differential amplifier). When the headlamps are turned on, the comparitor reads the input voltage. If the headlamp is normal there is a slight drop in voltage across the fixed resistor. But if the circuit opens, there is no change in voltage drop and the comparitor signals

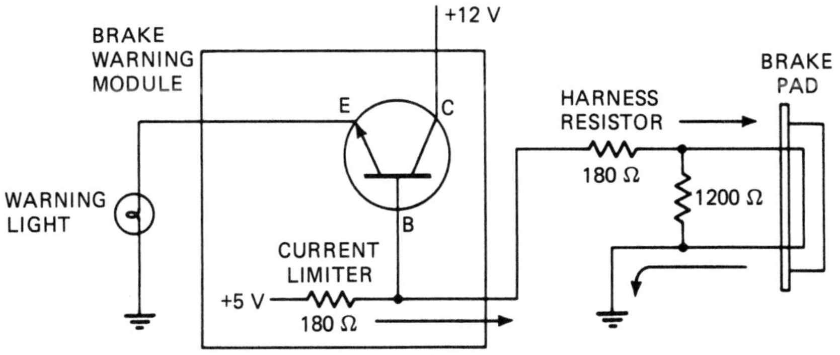

(B)

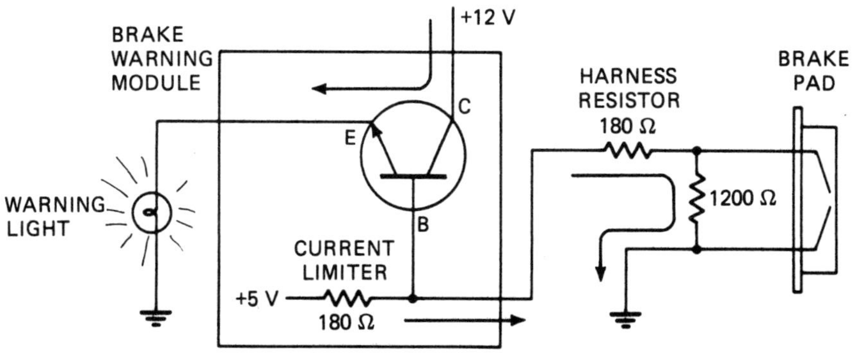

(C)

Figure 4-21 (continued) (b) Brake pad normal. Here is a more detailed version of how a brake warning system can work. With the transistor, the base must go positive for the transistor to conduct (turn on). Current can flow from the 5-volt source and bypass the 1,200Ω resistor. The result is the current limiter resistor (180Ω) is in series with the harness resistor (180Ω). The two equal resistors in series (see Chapter 1) will cause the applied voltage (5v) to be divided in half, providing 2.5 volts on the base of the transistor. Since 2.5 volts is not sufficient to turn the transistor on, the warning light remains off. (c) Brake pad worn. With the wire worn through in the brake pad, current now has to flow through the 1200Ω resistor. The added resistance in this series circuit will result in a voltage drop across the current limiter of approximately .6 volts. This will leave a remainder of +4.4 volts on the transistor base. This base voltage is sufficiently positive to turn the transistor on (make it conduct from collector to emitter) and light the warning light.

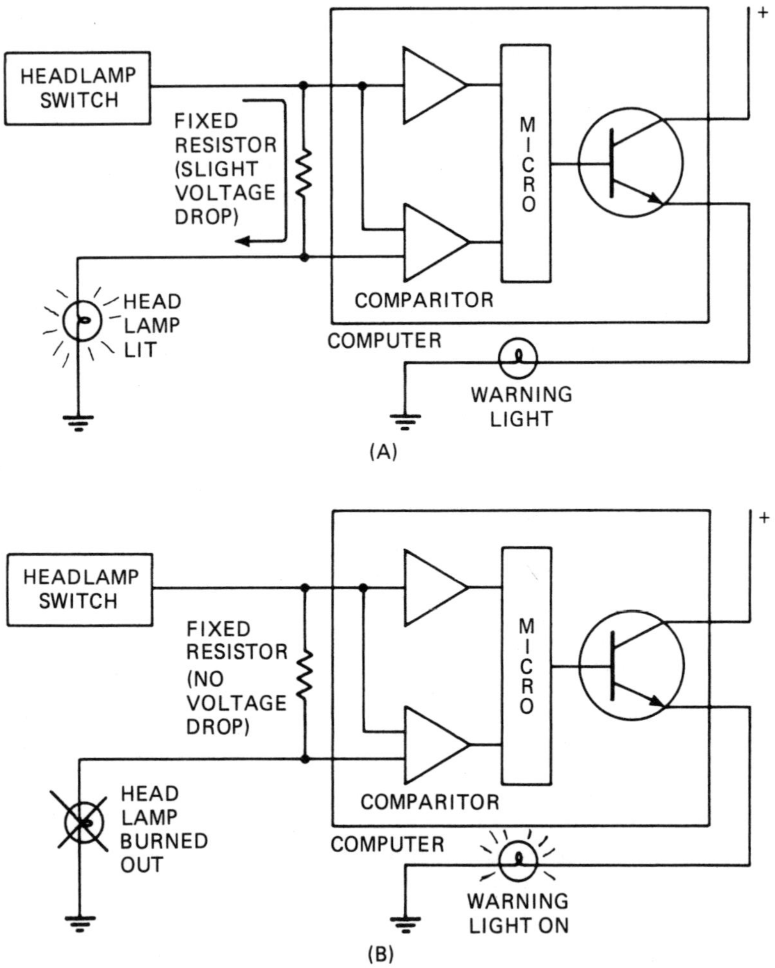

Figure 4-22 Headlamp warning indicator utilizing a comparitor. (a) Headlamp normal. (b) Headlamp burned out.

the microprocessor that a bulb has burned out. The sensing resistors are calibrated for each type of light bulb.

A comparitor can also be used to monitor a light circuit with more than one bulb connected in parallel (see Figure 4-23). If a bulb burned out, current flow would decrease and the voltage drop would change across the fixed resistor (See Figure 4-23b). The comparitor would signal the microprocessor of the change and the microprocessor would turn on a warning light indicating a failure in the circuit.

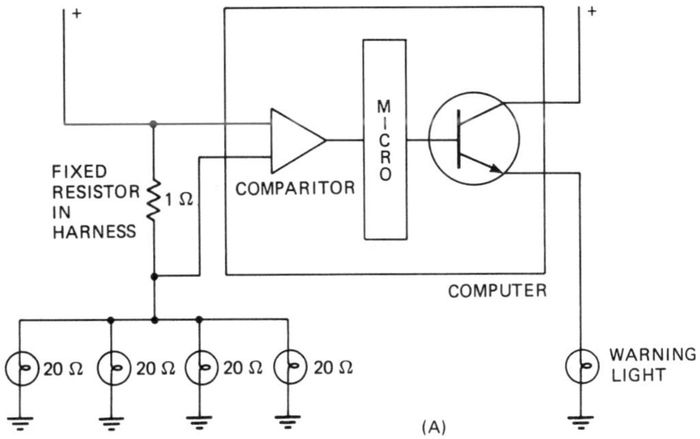

Figure 4-23 (a) Warning indicator for four lamps utilizing a comparitor. All lamps normal. As explained in Chapter 1, total resistance of the four lights in parallel is

$$\frac{20\Omega}{4(\text{lamps})} = 5\Omega$$

The total circuit resistance is $5\Omega + 1\Omega$ (fixed resistor) $= 6\Omega$. The total current ($I = E/R$) is

$$I = \frac{12V}{6\Omega} = 2A$$

Voltage drop across the fixed resistor ($E = I \times R$) is

$$E = 2A \times 1\Omega = 2 \text{ volts}$$

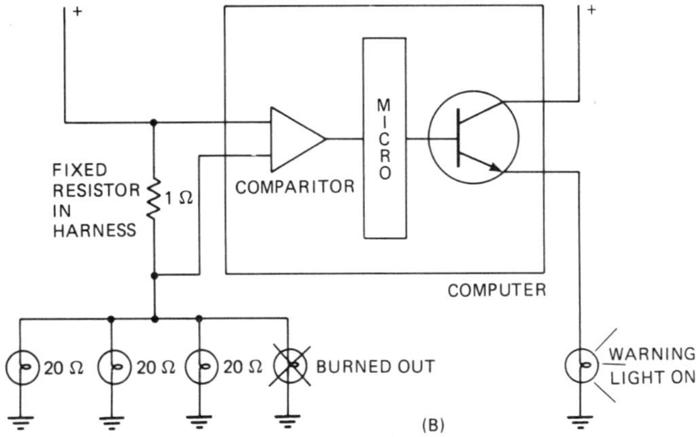

Figure 4-23 (b) With one bulb out, total resistance of the three remaining lamps is

$$\frac{20\Omega}{3(\text{lamps})} = 6.66\Omega$$

The total circuit resistance is $5\Omega + 1\Omega$ (fixed resistor) $= 6\Omega$. Current flow ($I = E/R$) is

$$I = \frac{12V}{7.66\Omega} = 1.57A$$

Voltage drop across the fixed resistor ($E = I \times R$) now is

$$E = 1.57A \times 1\Omega = 1.57 \text{ volts}$$

The microprocessor senses the reduction in voltage drop across the fixed resistor (from 4-23a) and lights the warning light.

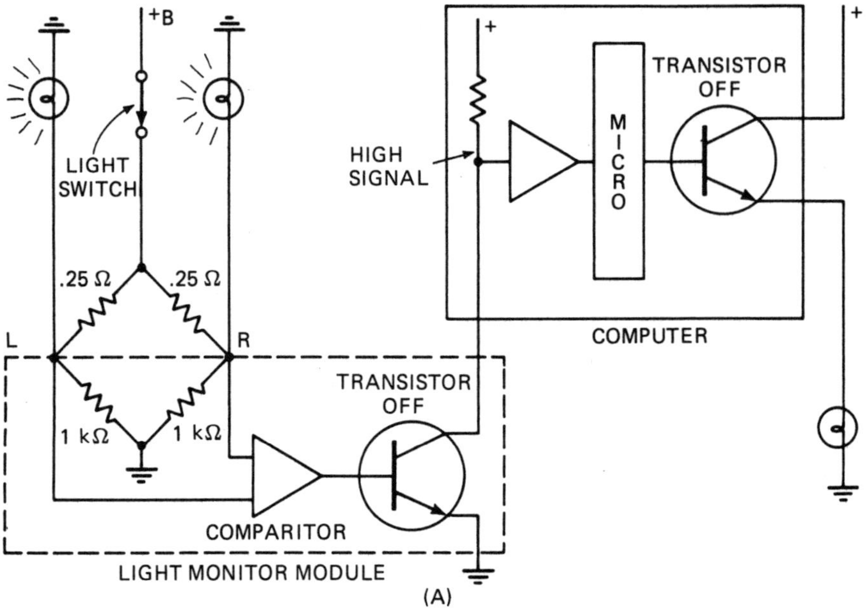

Figure 4-24 Warning indicator for headlamps utilizing Wheatstone bridge circuit. (a) Headlamps normal. With resistances as shown, voltages at points "L" and "R" on the Wheatstone bridge would be 11.3 volts. Equal voltages to both sides of the comparitor would turn the transistor off. The resulting high voltage into the microprocessor would leave the warning light off.

The second type uses a Wheatstone bridge circuit to detect changes in resistance for each side of a common light circuit (such as the left and right low beam headlights, see Figure 4-24a). Current flows through separate fixed resistors which form the top of the Wheatstone bridge. Most of the current will then flow out to the lights. Inside the computer are two high resistance fixed resistors which form the bottom of the Wheatstone bridge. A comparitor (differential amplifier) will measure the voltage drop after the top pair of resistors. When the circuits are normal, the bridge is balanced and the voltages received by the comparitor are equal. But if a bulb burns out (see Figure 4-24b), the bridge becomes unbalanced. The comparitor reads the change in voltage and turns on a transistor. The transistor is wired in series with a fixed resistor and the computer monitors the voltage drop across the transistor. With the

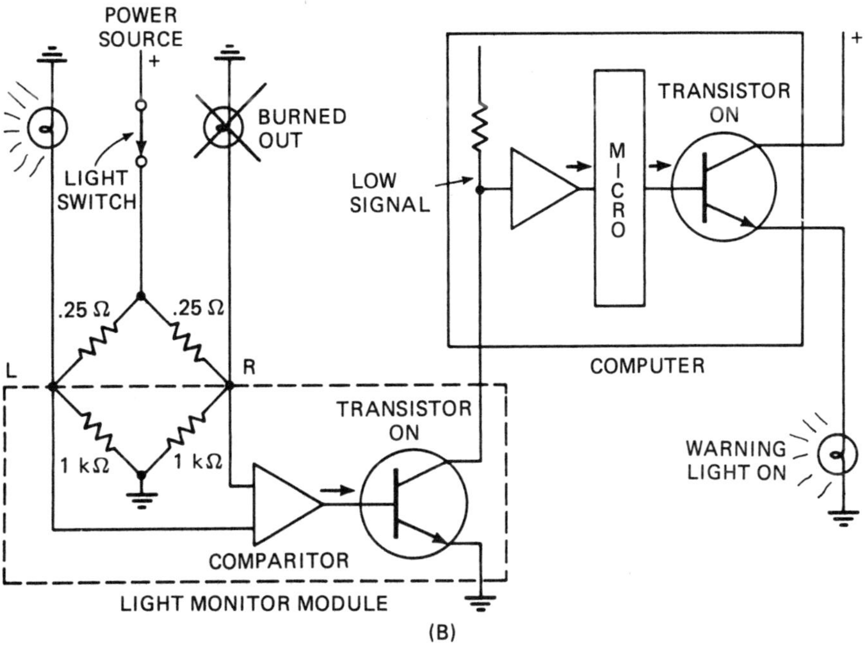

Figure 4-24 (continued) (b) One headlamp burned out. With the right light out, voltage at point "L" would still be 11.3 volts, but voltage at point "R" would now be 11.99 volts. The voltage difference between points "L" and "R" would cause the comparitor to turn the transistor on. The low signal input to the microprocessor would result in the warning light being turned on.

transistor turned on the voltage drop is low and the computer will turn on a warning light indicating a bulb has burned out.

The number of Wheatstone circuits used depends on the number of light circuits the manufacturer wants to have monitored (see Figure 4-25).

4-1-8 Function Switches

Function switches are simply on/off digital switches. These switches command the computer and provide certain information to the driver. This information could be about such items as converting English units to metric units or the number of gallons left in the fuel tank. Regardless of the function, the circuits all work in the same manner. The switch is located on the ground side of a fixed

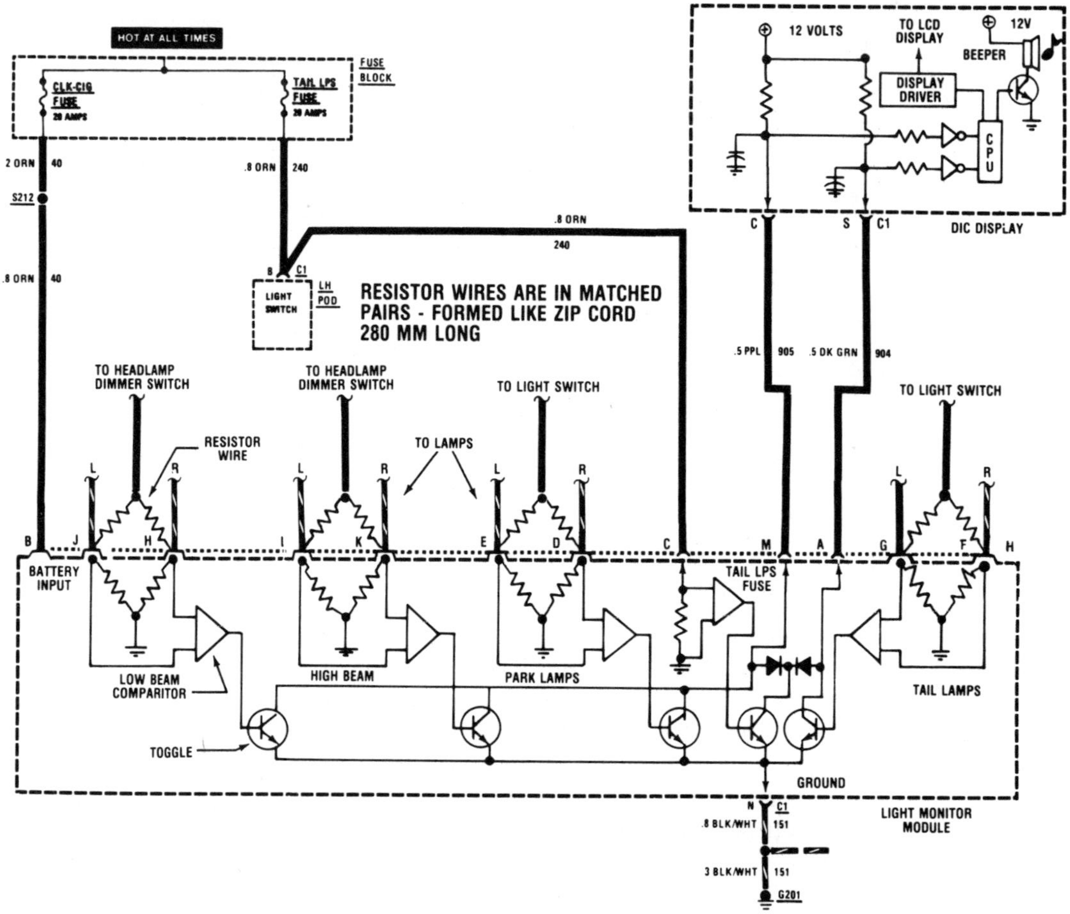

Figure 4-25 Wheatstone bridge lamp-out monitoring system. (Courtesy of General Motors Product Service Training.)

resistor (see Figure 4-26). When the switch is pressed, the circuit is completed and the computer recognizes the change in voltage and knows the driver now wants that particular function displayed.

§ 4-2 ELECTRONIC INSTRUMENT COMPUTER

The electronic instrument computer is very similar to the engine computer. It contains many of the same components as the engine computer and therefore that information will not be repeated. Its main difference, however, is in its function and programming. It

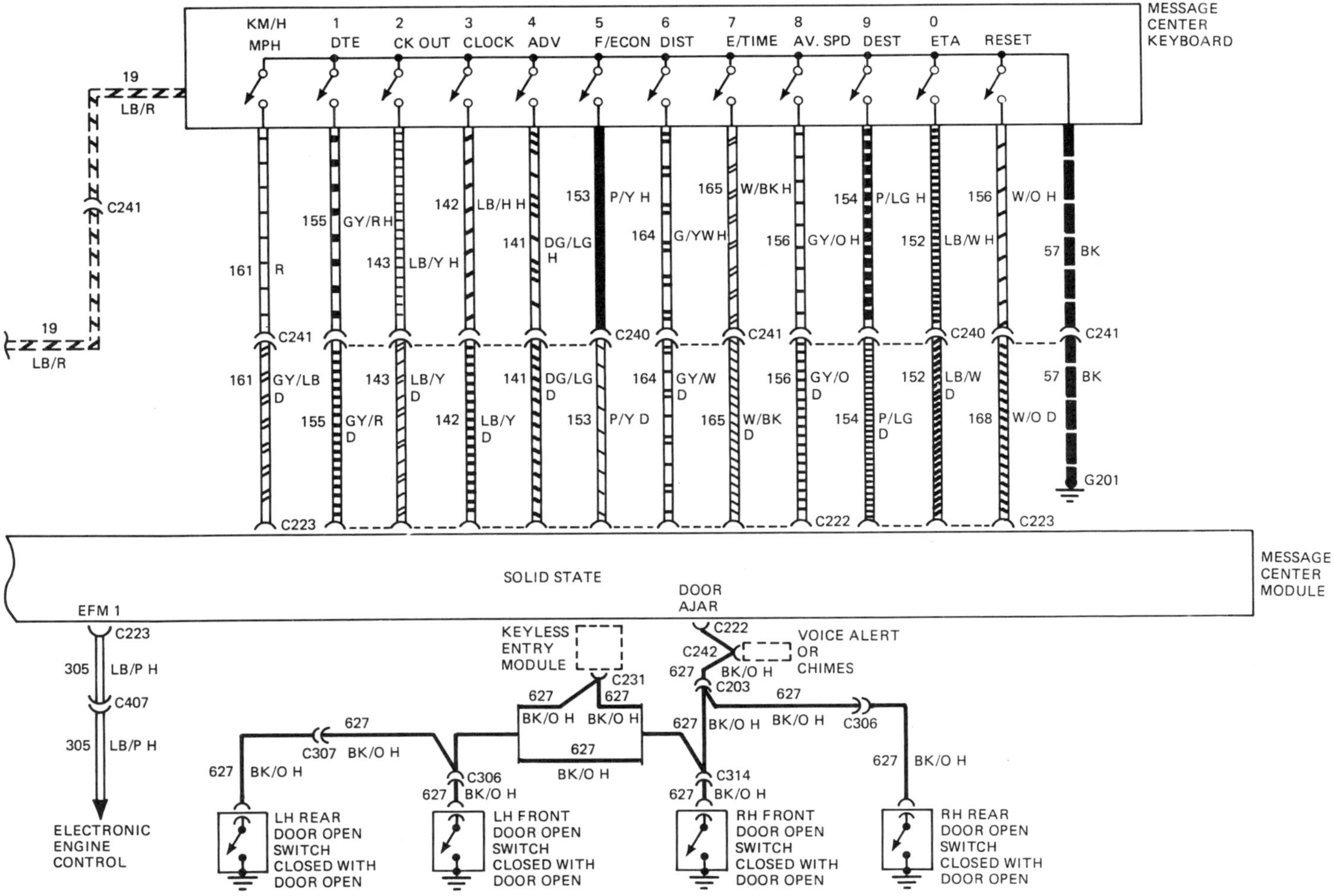

Figure 4-26 Typical function switches. (Courtesy of Ford Motor Company.)

must interpret and turn on the proper display telling the driver what is happening. Its programming depends on what it was designed to do. Through programming, the computer can calculate and present such information as miles traveled, number of gallons of fuel used, the time, estimated time of arrival—and the list continues to grow with each new model year. How this information is presented is also changing; this topic is discussed in the next section.

§ 4-3 DIGITAL DISPLAYS

Digital displays are composed of common light bulbs, light emitting diodes (LED), vacuum tube fluorescent (VTF), and liquid crystal displays (LCD). Each of these displays have different characteristics and operate differently.

An LED uses a semiconductor element that glows when current passes through it (see Figure 4-27). A plastic lens is used to diffuse the light. The color depends on the construction of the LED. LED's come in different colors such as red, yellow, or green. LED's are connected together to create numbers or letters, Figure 4-28. A major disadvantage of LED's is that they have a high power consumption versus light output. This usually means that when the dash is in sunlight, the LED is hard to see.

VTF's are currently the most common form of lighting a digital dash. They are generally recognizable by their blue-green color. There are three major components contained within a glass ampule

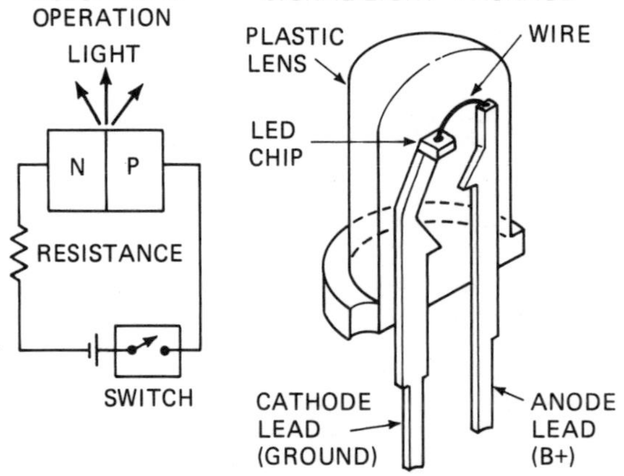

Figure 4-27 Light emitting diode (LED). (Courtesy of Ford Motor Company.)

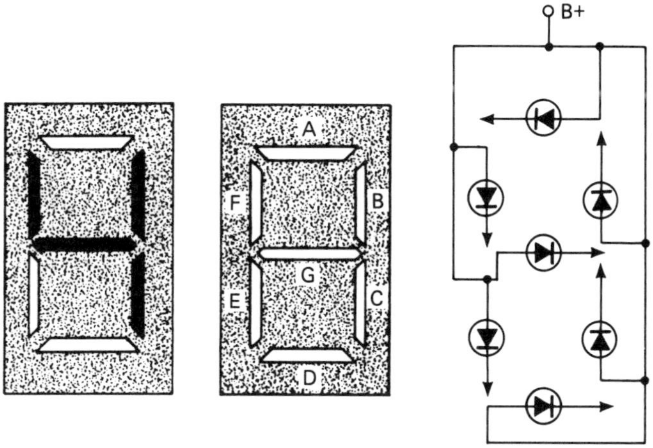

Figure 4-28 A typical LED display. (Courtesy of General Motors Product Service Training.)

filled with either argon or neon gas in the VTF: a tungsten filament heater, a grid, and a phosphor-coated anode (see Figure 4-29a). The tungsten filament heater, which is heated by a low voltage current, gives off a cloud of electrons. The grid uses a higher voltage pulsating current; it attracts and accelerates the electrons toward the anode. The anode, using a more rapid pulsating current, attracts the electrons. As the electrons strike the phosphor coating of the anode, light is given off (see Figure 4-29b). The computer will apply voltage to the bar segments it wants lit to represent the correct alpha-numeric character (see Figure 4-30). VTF's are very bright and are often too bright for nighttime use. Usually the dash VTF's are connected to the headlamp circuit. When the headlights are turned on, voltage to the anode is reduced and less light is given off. Another method uses a photosensor to detect ambient light. As the ambient light level increases, the brightness of the dash increases.

LCD's make up the newer and more sophisticated digital dashes. They can be constructed in a variety of shapes and colors. A major difference between the LCD and LED is that the LCD does not give off its own light. It must use ambient light or an external light source. A special fluid is used which has the ability to change light wave polarity. This fluid is a crystal in liquid form. The transmission or reflection of light is affected by applying a potential dif-

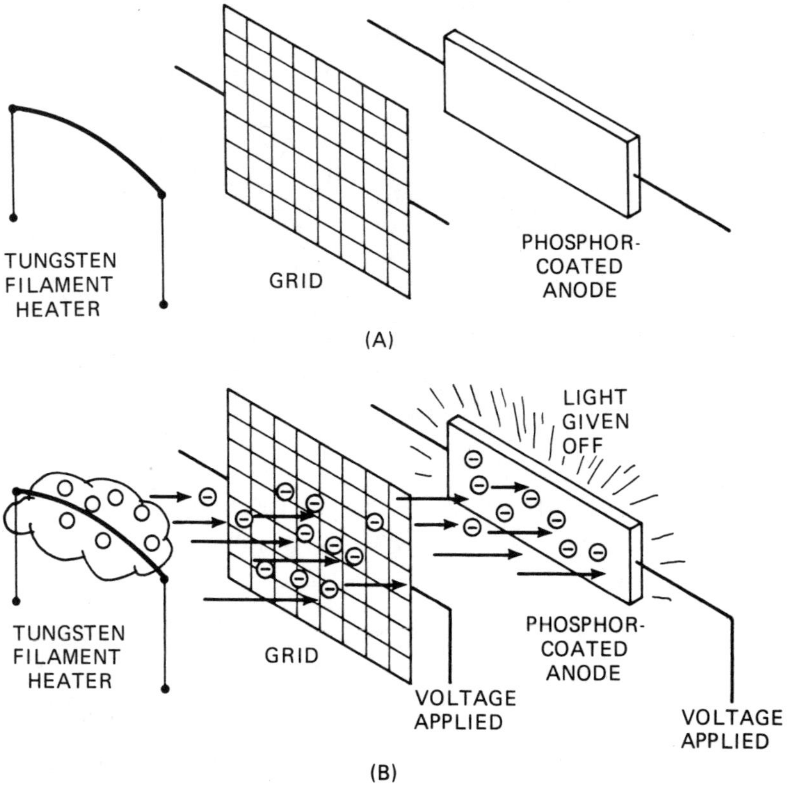

Figure 4-29 Vacuum tube fluorescent (VTF). (a) Display components. (b) Electrons emitted from the VTF heater are accelerated by the grid voltage and are attracted to the phospher-coated anode.

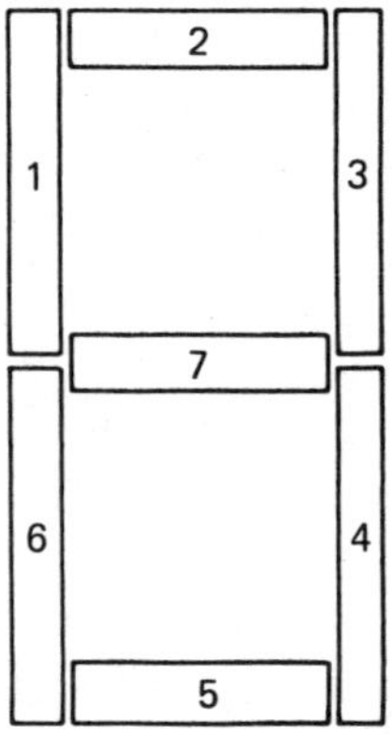

Figure 4-30 Bar segments of a vacuum tube fluorescent (VTF) display.

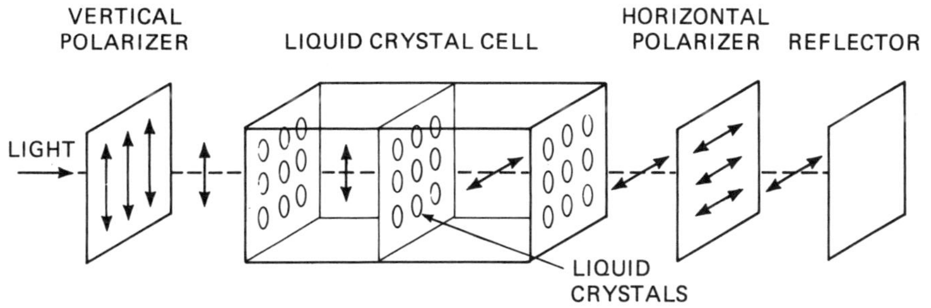

Figure 4-31 Reflective LCD operation with no voltage applied. (Courtesy of Ford Motor Company.)

ference to the liquid crystal. There are basically two type of LCD's: reflective and transmissive.

The reflective type is commonly used in wristwatches, calculators, and test equipment such as digital volt/ohm meters and uses ambient light as its light source. It is made of a vertical polarizer, liquid crystal cell, horizontal polarizer, and reflector (see Figure 4-31). The vertical polarizer allows only vertical light waves to pass through. At this time there is no voltage applied to the crystal and the light wave is gradually rotated 180 degrees, where it passes through the horizontal polarizer to the reflector. The light waves bounce off the reflector, through the horizontal polarizer, liquid crystal, and vertical polarizer. The viewer will see only reflected light. But when a voltage is applied to a segment of the crystal, the liquid crystal molecules in the fluid assume a symmetrical pattern. The vertical light waves will no longer rotate and cannot pass through the horizontal polarizer to the reflector (see Figure 4-32). Since no

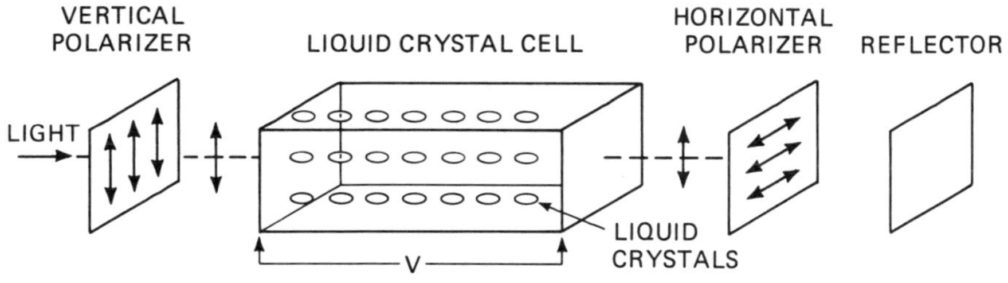

Figure 4-32 Reflective LCD operation with voltage applied. (Courtesy of Ford Motor Company.)

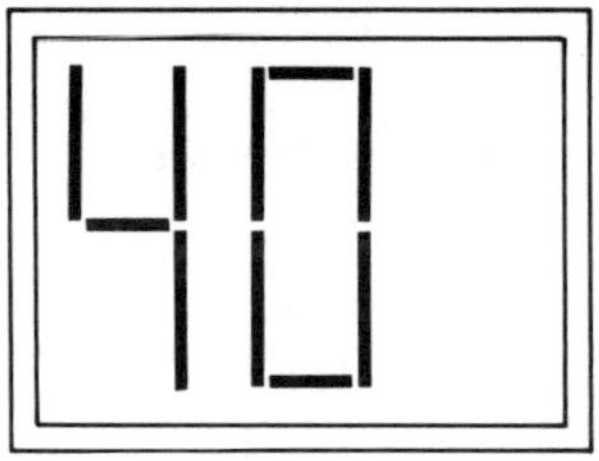

Figure 4-33 LCD display. Ten segments had voltage applied to produce the number 40.

light is reflected in this portion, the segment appears dark to the viewer (see Figure 4-33).

The other type of LCD is the transmissive type, which uses an artificial light source on one side of the LCD. The liquid crystal is contained between two pieces of polarized glass (see Figure 4-34). Each piece of glass has a special electric conductive surface laminated to it. When a square wave voltage of equal phase and intensity is applied to each glass, there is no difference in potential between the two pieces of glass. Light cannot pass through, for the liquid crystal molecules are in a twisted, random state. But when a segment on one piece of glass has its voltage shifted out of phase (see Figure 4-34b), one glass plate will have a high voltage, the other a low voltage, creating a difference in voltages. The liquid crystal

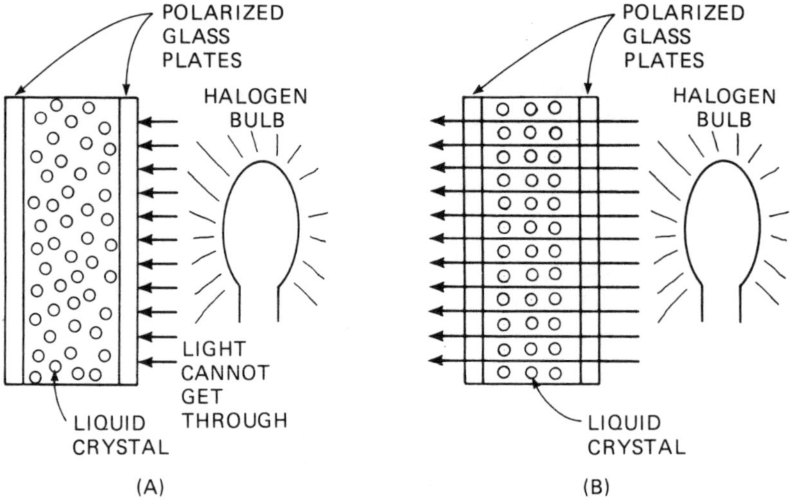

Figure 4-34 Transmissive LCD. (a) No voltage applied. (b) Voltage applied.

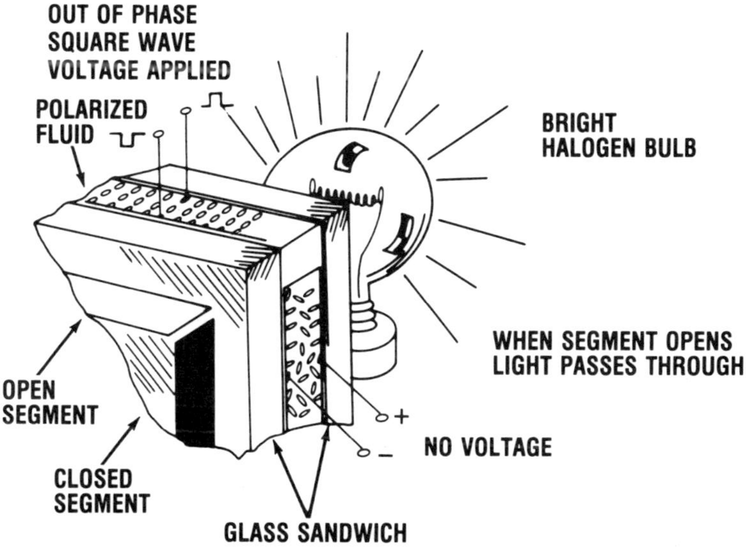

Figure 4-35 Transmissive LCD. (Courtesy of General Motors Product Service Training.)

molecules align themselves and this allows the passage of polarized light through (see Figure 4-35).

§ 4-4 ANTILOCK (SKID) BRAKE SYSTEMS

The antilock brake system's purpose is to prevent the wheels from locking up regardless of the condition of the road surface. With antilock brakes, it is possible to stop in a straight line or maintain steering capability on slippery surfaces.

The hydraulic and mechanical brake system operation and parts are essentially the same. What is different is that a computer will control brake pressure to the wheels, preventing lockup.

Variable reluctance-type sensors are located at each wheel (the exception is to use a variable reluctance sensor on the drive shaft or on the differential ring gear to sense what the rear wheels are doing). A toothed sensor ring is mounted on the rotor, axle shaft, or drive shaft (see Figure 4-36). As each tooth passes by a sensor assembly, consisting of a permanent magnet and coil, an AC voltage is induced in the coil. A buffered circuit is used to convert the AC signal

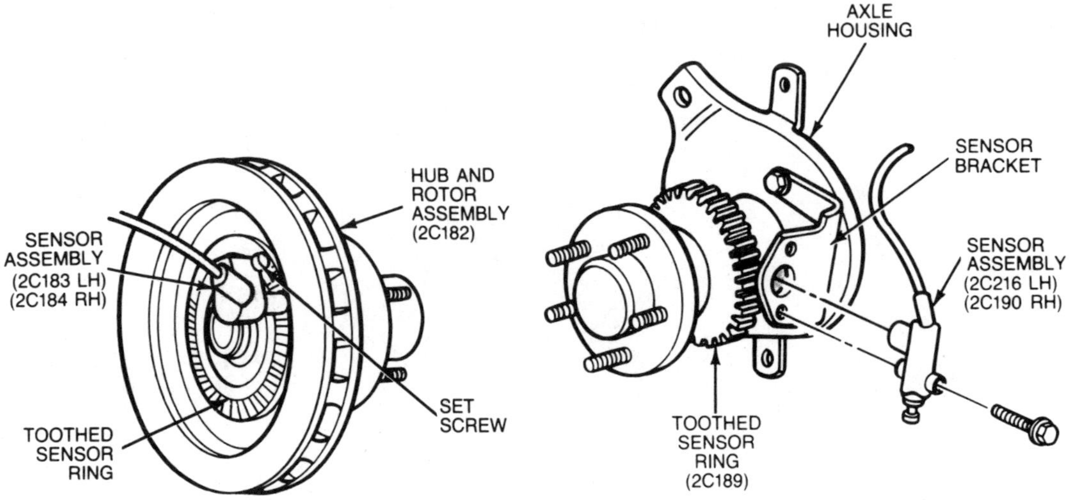

Figure 4-36 Typical antilock brake wheel type sensors. (Courtesy of Ford Motor Company.)

to a digital signal (note that this circuit is nearly identical to the PM buffered speedometer circuit). The frequency of this signal is directly proportional to wheel speed. From the signal frequency, the computer is able to determine wheel speed and rate of deceleration. For when lockup is about to occur, signal frequency decreases at a rapid rate. Then the computer takes over, modulating brake pressure to the wheels.

There are two different methods for modulating brake pressure, though they both use solenoids.

One type uses normally open and normally closed solenoids (see Figure 4-37). Each front wheel brake line has one normally open and one closed solenoid. The rear wheels are connected to a common brake line and therefore share one pair of normally open and closed solenoids. The operation for each wheel is the same. Brake pressure under normal conditions is allowed to go to the wheel. In this mode the normally open solenoid is open and the normally closed solenoid is closed. But when lockup is about to occur, the computer will close the open solenoid and open the normally closed solenoid, reducing brake pressure. The signal frequency of the wheel will then increase and allow the solenoids to go back to their original positions. This process is cycled very rapidly. The driver

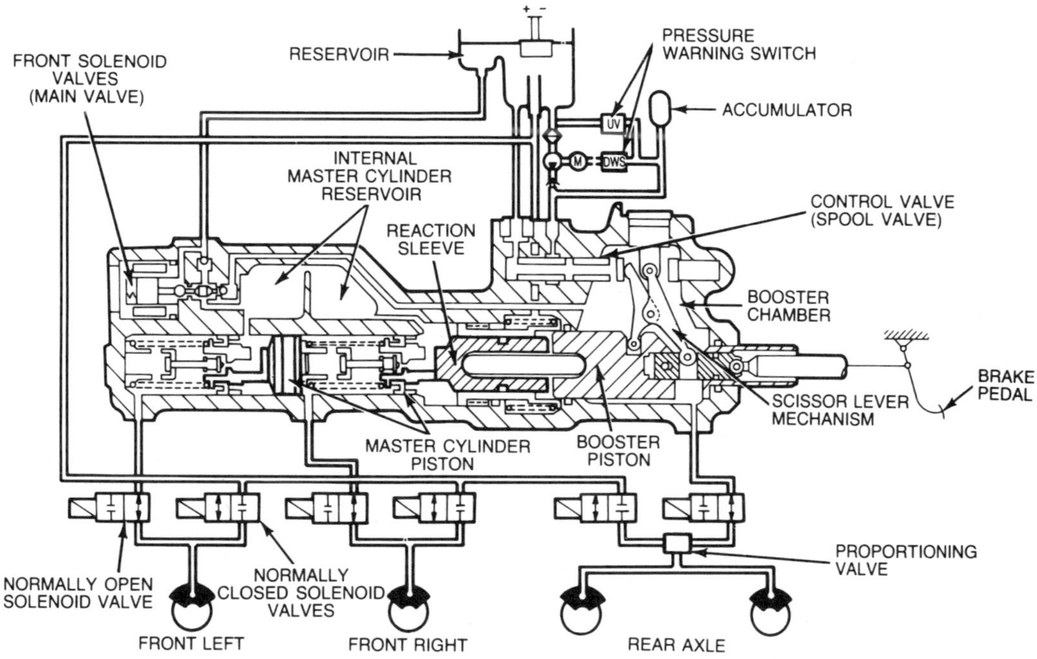

Figure 4-37 TEVES antilock brake system. (Courtesy of Ford Motor Company.)

may feel pulsations in the brake pedal under these conditions, which is normal.

Brake pressure is supplied by the driver and assisted by an electric pump or vacuum booster. But there must be a means to supply brake pressure while in the antiskid mode, so that the brake pedal does not sink to the floor. By using an electric over hydraulic pump, pressure is supplied on demand.

Another type uses a normally open solenoid for each brake circuit and a hydraulic modulator valve. During normal braking the solenoids are open and allow brake pressure to go to each wheel (see Figure 4-38). When the computer senses that lockup is about to occur, the solenoid is only partially energized, preventing additional pressure or fluid to reach the wheel (see Figure 4-39). If the computer calculates that lockup will occur, then it will fully energize the solenoid. This moves the valve into a position where brake pressure from the caliper will be reduced (see Figure 4-40). Brake fluid will be stored in an accumulator. Also, the return pump is activated,

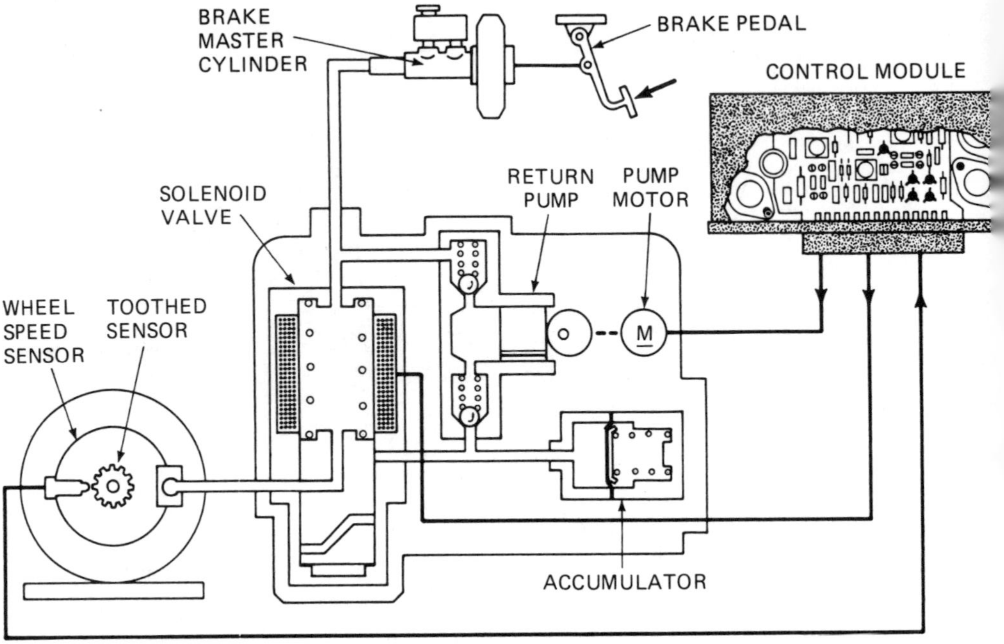

Figure 4-38 Bosch antilock system. Solenoid in normal position. (Courtesy of General Motors Product Service Training.)

sending fluid back to the master cylinder against brake pedal pressure. The owner will feel rapid pulsations in the brake pedal. This is normal.

The computer for antilock brakes will supply only trouble codes related to its circuitry and sensors. It will not diagnose hydraulic or mechanical brake failures. A separate indicator light is used to indicate if the antilock system is inoperative. Normal braking should still occur. Always consult the service manual when working on antilock brakes. Some portions of the system will store brake fluid under high pressure. Sudden release of this pressure can injure you and/or do extensive damage to the car.

§ 4-5 AIR SUSPENSION

In place of a mechanical spring an air-filled bladder is used to suspend the vehicle body (see Figure 4-41). Air makes an ideal spring, for its spring rate increases as it is compressed. Another ad-

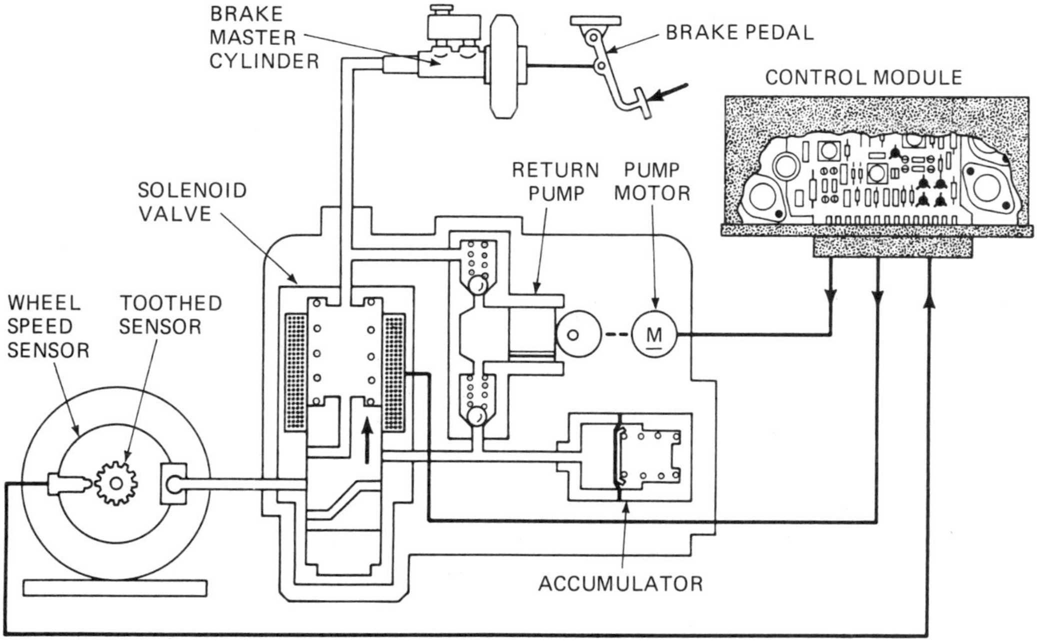

Figure 4-39 Solenoid valve partially energized, blocking brake pedal pressure to the wheel. (Courtesy of General Motors Product Service Training.)

vantage is that when connected to a computer controlled air compressor, vehicle riding height can be maintained under a variety of loads.

The inputs consist of riding height sensors—one located at each front wheel and one at the rear of the vehicle (see Figure 4-42). Another input is an on/off switch which is used to turn the system off in some conditions.

The ride height sensor consists of a slide with magnets mounted on it. Two Hall effect sensors, one at the top, the other at the bottom, are mounted in the slide housing (see Figure 4-43). When the vehicle is at the proper height, both switches are closed. But if riding height should decrease, one switch will open, indicating the vehicle is out of trim. When the vehicle comes back to its trim height, both switches are closed again.

The computer controls two major outputs: the air compressor and a normally closed solenoid on each air spring. If vehicle height needs to be raised, the computer will turn on the air compressor and

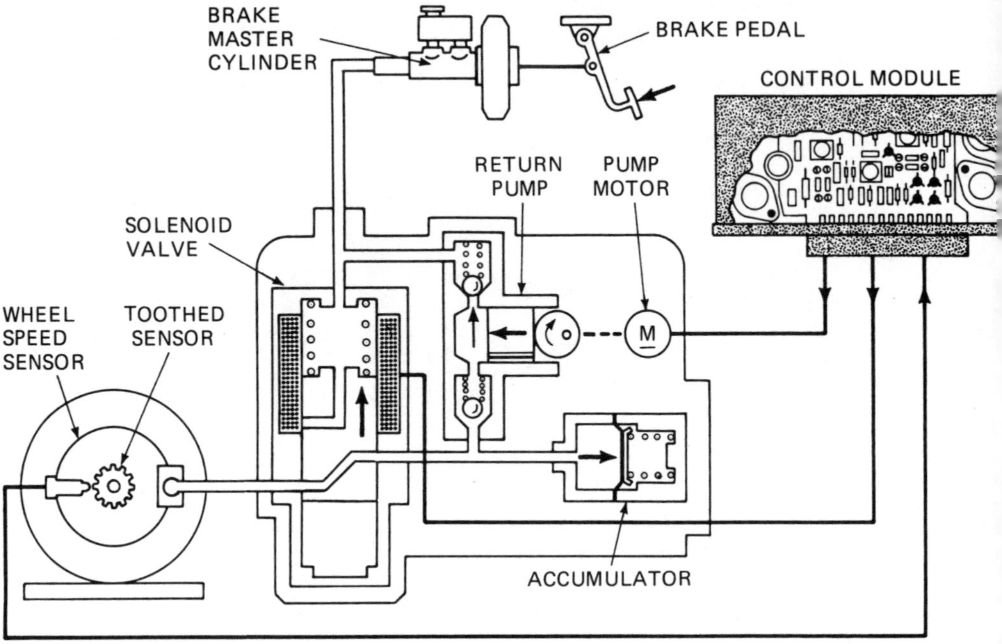

Figure 4-40 Solenoid valve fully energized, venting trapped brake fluid and decreasing system pressure. (Courtesy of General Motors Product Service Training.)

pump outside air through a dryer to the springs. To lower vehicle height, the computer can individually energize the proper solenoid and release trapped air pressure (see Figure 4-44). The computer will also open a system vent valve and allow the air to pass back through the dryer and remove any moisture that had been accumulated when the air had entered the system.

The air suspension system must be turned off when the vehicle is raised or towed. If not, the computer would try to compensate. For example, if the car is raised on the hoist with the wheels suspended, the computer will think that the riding height has increased and vent pressure out of the air springs. When the vehicle is put back on the ground the suspension will have bottomed out and the springs may be damaged. The vehicle owners manual will give the location of the on-off switch (usually in the trunk).

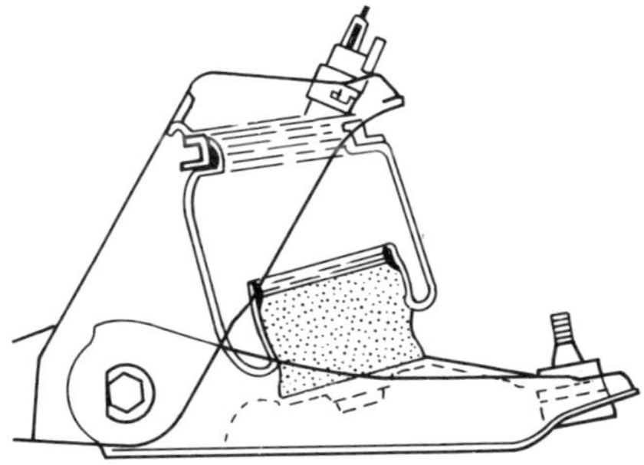

(A)

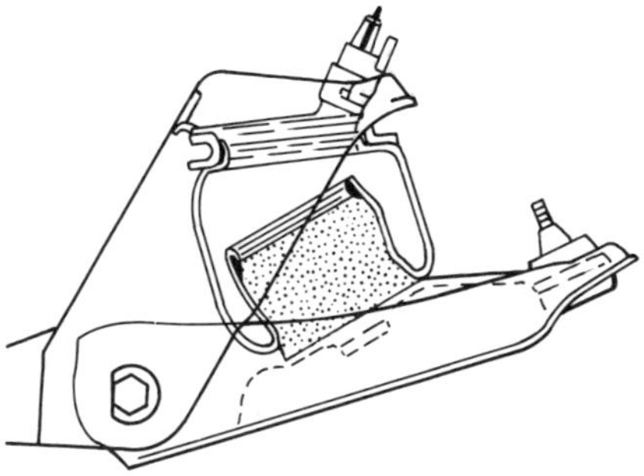

(B)

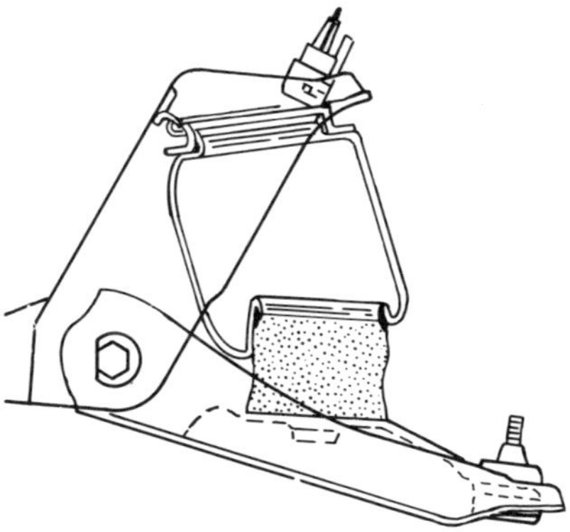

(C)

Figure 4-41 Air spring. (a) At normal height. Air pressure contined in the rubber membrane maintains vehicle height and acts like a coil spring. The air spring valve mounted in the end cap opens to allow air to enter and exit spring. When air is added, the vehicle will rise; when air is removed, the vehicle will lower. (b) Jounce. When control arm moves upward, the piston moves upward into rubber membrane. As the arm moves upward toward jounce, the rate of the air spring increases. (c) Rebound. When control arm moves downward, piston extends outward from rubber membrane. The rubber membrane unfolds from around the piston to allow downward suspension movement.

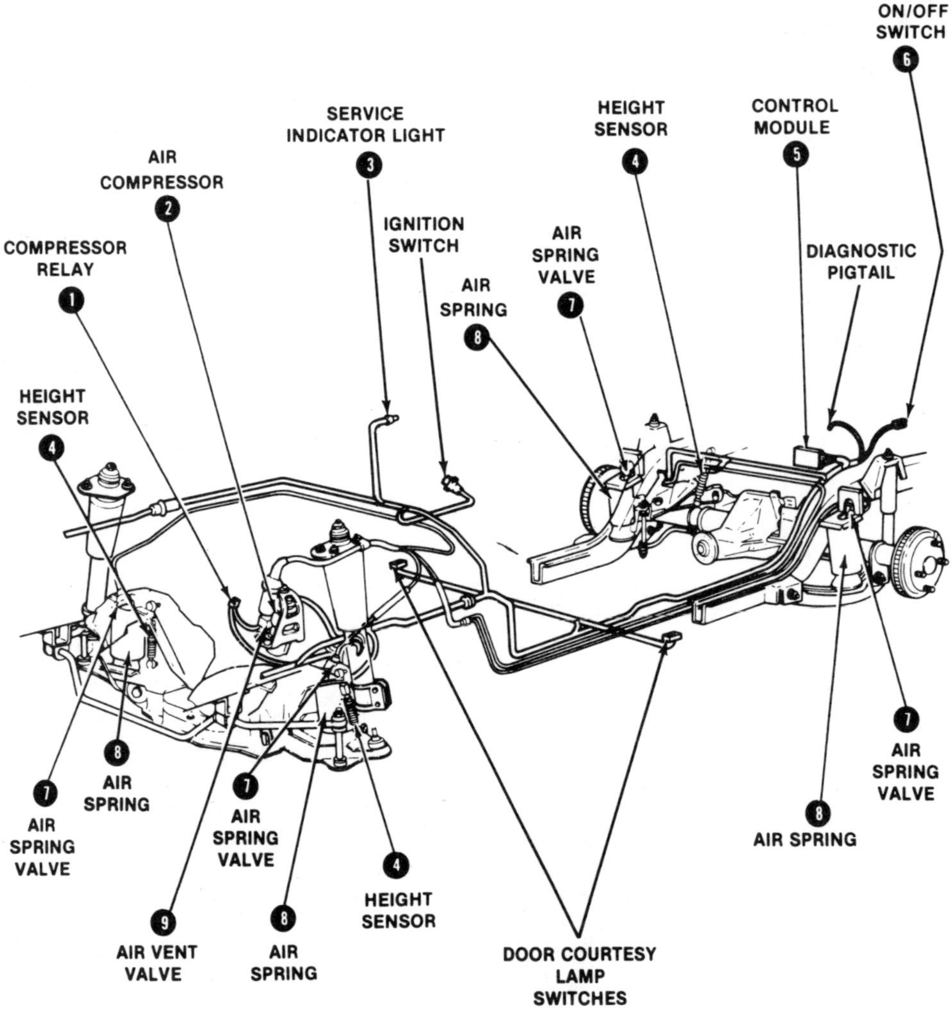

Figure 4-42 Air suspension system components. (Courtesy of Ford Motor Company.)

§ 4-6 CRUISE CONTROL

The cruise control system enables the driver to set the vehicle speed at a desired setting and have that speed maintained under normal highway conditions. The system can be activated above approximately 30 miles per hour and will operate the throttle as neces-

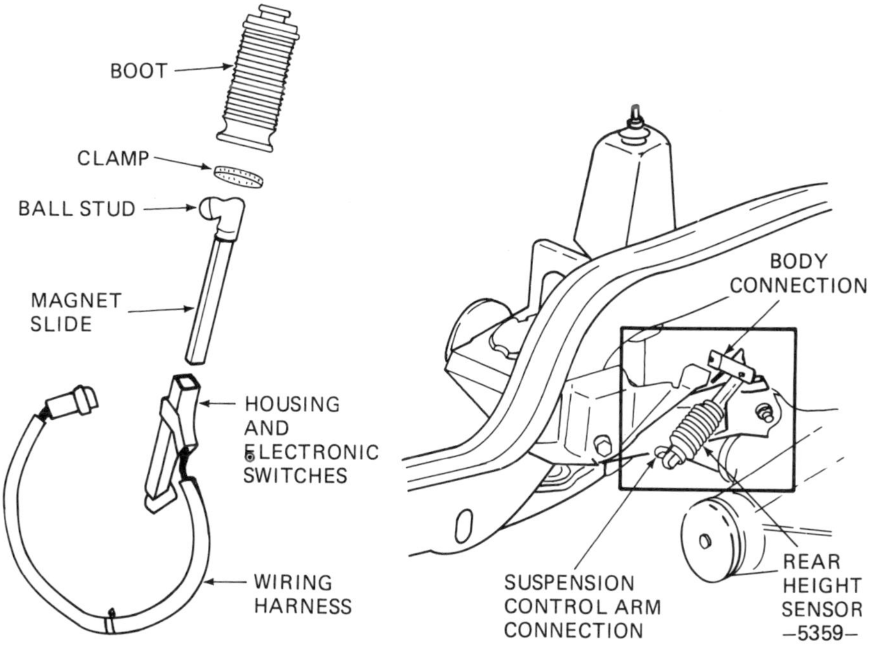

Figure 4-43 Height sensor. (Courtesy of Ford Motor Company.)

sary to maintain the set speed. Vehicle braking is *not* provided, so the vehicle can gain speed when descending steep grades.

The main parts of the system are the throttle actuator (usually vacuum operated), the amplifier (which may be incorporated into the engine computer), the speed control switches, and the speed sensor (see Figure 4-45).

The speed sensor sends a signal to the amplifier that changes in frequency and amplitude with vehicle speed. When the driver turns on the cruise control by setting vehicle speed, the amplifier will sample the sensor output at that speed. From then on it will try to keep the speed sensor output the same as the sample output. It controls the speed sensor output by controlling vehicle speed with the throttle actuator. If the vehicle slows down, the sensor output drops below the sample output. The amplifier will then operate the throttle actuator to open the throttle, increasing vehicle speed and speed sensor output until the sensor output matches the sample output.

If the driver steps on the brake, output from the stoplight circuit will signal the amplifier to disengage the cruise control. Most

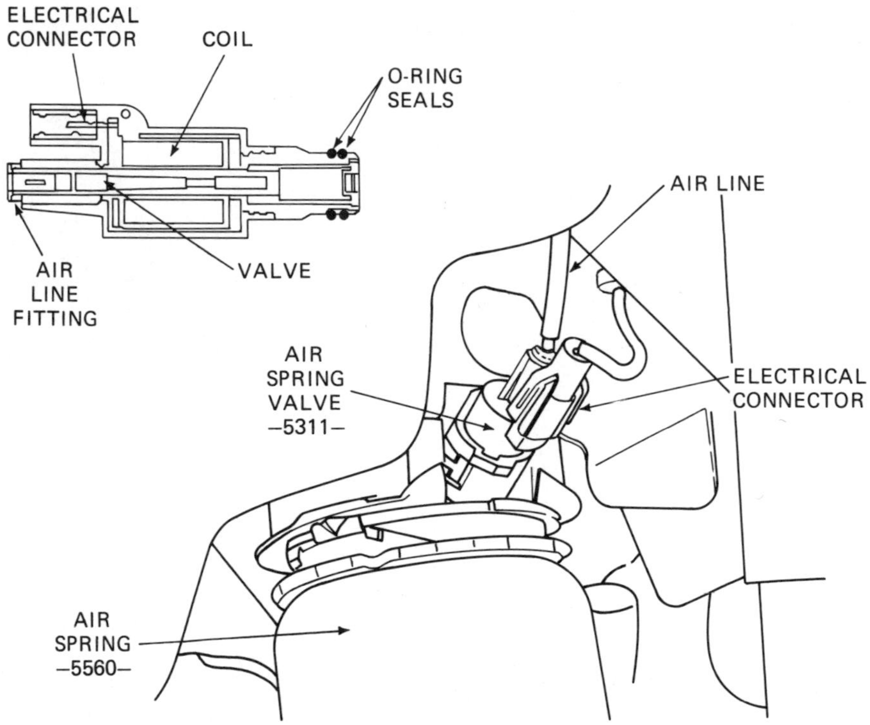

Figure 4-44 Air spring valve. (Courtesy of Ford Motor Company.)

systems have a resume feature, which means that the sample output is maintained in the amplifier (as long as the ignition switch remains on). This enables the driver to go back to the previously set speed by simply pressing the resume button. This locks the system back on and allows the actuator to again try to match speed sensor output to sample output.

To provide smooth operation and prevent the system from hunting above and below the set speed, a potentiometer is connected to the throttle actuator. This signals the amplifier as to the actual throttle position so the amplifier can quickly determine how much throttle correction is needed in the event that vehicle speed should vary from the set speed.

The amplifier controls the actuator by operating a pair of solenoid valves which regulate vacuum or vent air into the actuator. Figure 4-46 shows a typical system.

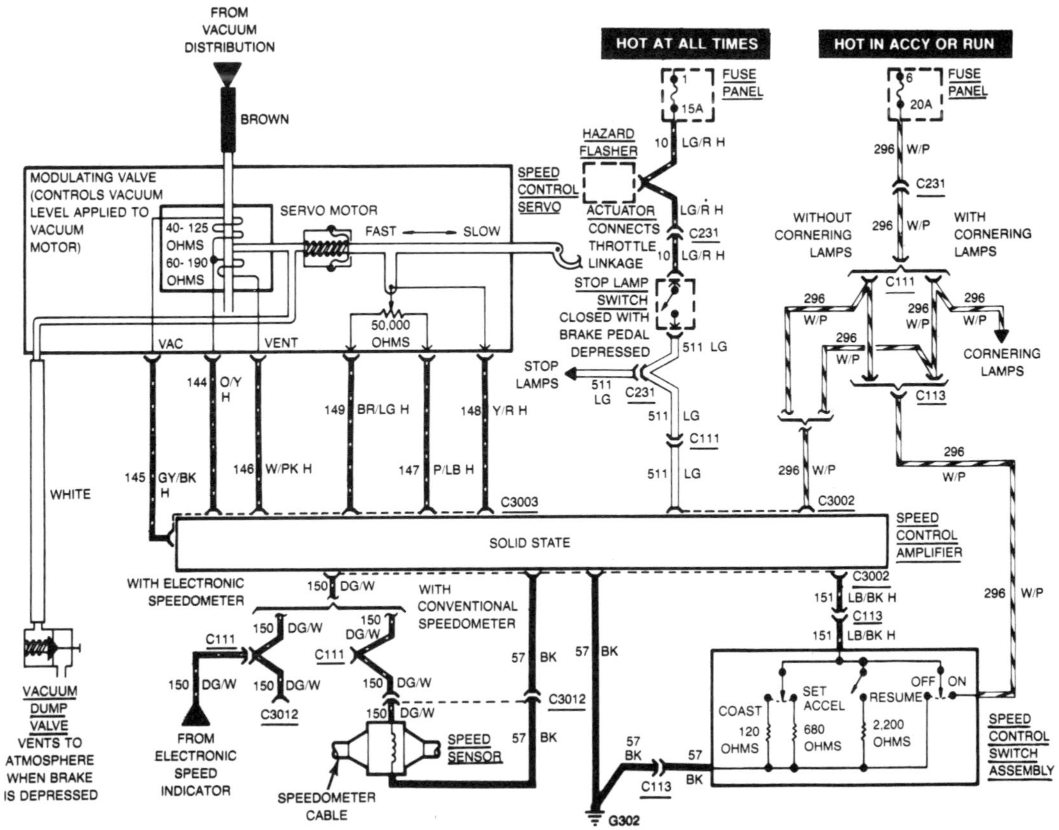

Figure 4-45 Typical speed control electrical schematic. (Courtesy of Ford Motor Company.)

QUESTIONS

4-1. Describe what electronic instrumentation must accomplish.

4-2. Name two different types of speedometer sensors and describe how they work.

4-3. Describe how temperature sensing circuits work.

4-4. Name the two different types of oil pressure sensors and describe how the circuit works.

4-5. How does the fuel level circuit work?

4-6. What sensors are used for a tachometer?

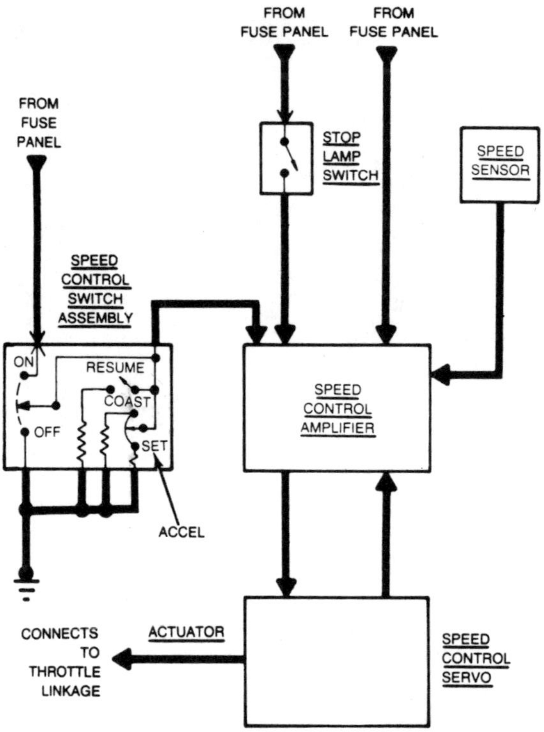

Figure 4-46 Speed control system. (Courtesy of Ford Motor Company.)

4-7. Describe how a comparator warning circuit works.

4-8. Describe how a Wheatstone bridge warning circuit works.

4-9. What is the major disadvantage of the LED?

4-10. Describe how VTF works.

4-11. Describe how a reflective LCD works, and name some common applications.

4-12. Describe how a transmissive LCD operates. Where is it commonly used?

4-13. Name the major inputs to the antilock breaking system. Describe their function.

4-14. How does the computer control brake line pressures?

Glossary

Actuator A device that delivers motion in response to an electrical signal.

A/D converter (signal processor) Analog/Digital Converter: A circuit within the computer that takes an analog signal and converts it to a digital signal that can be used by the microprocessor.

Alpha numeric Display capable of representing numbers or letters on the alphabet.

Alternating current (AC) Electrical flow that periodically reverses direction. One flow of current in each direction is called a "cycle."

Ampere A unit of measurement for the flow of electric current. One ampere is the rate of electron flow when one volt of pressure is applied across one ohm of resistance.

Amplifier A device used to increased the strength of an electrical signal.

Amplitude The maximum departure from its zero value of an electric current or voltage. In the case of the alternating current, it can be measured in either direction from zero.

Analog Electrical signal that varies continuously and smoothly over a certain range.

Anti-skid or anti-lock brake system A system that prevents the automobile brakes from "locking" in icy or panic situations. Reduces stopping distance and allows the driver to retain steering control.

Atom The smallest division of an element. The major parts of an atom are protons, neutrons, and electrons. Electric current flow is the movement of the electrons.

Binary (Binary system) A system of counting that can express any number by means of two digits, *0* and *1*. Applied to the operation of digital computers.

Bit The smallest possible piece of information. One out of two possible alternatives. Bits are written as *1* for *yes* and *0* for *no*.

Buffer A circuit used to separate the effects of two electronic circuits. In some systems it refers to a device that changes an A/C signal to a digital square-wave signal.

Byte A group of 8 bits.

Capacitance The property of a device or circuit part to store electrically separated changes. A capacitor is a device designed to store electrical changes.

Capacitor A device designed to store electrical changes.

Capacity capsule A device designed to change its capacitance by changes in the pressure of the medium that surrounds it.

Closed loop A system that regulates its output by feeding output information back into the input side of the regulating device.

Comparitor (Differential Amplifier) A device that produces an output voltage that is proportional to the differences between two input voltages.

Computer An automotive computer takes information from key points in the system and based on this information makes adjustments to the system to provide the desired operating characteristics.

Conductor A material that offers little opposition to the flow of electrons.

Conventional theory A practice of describing electrical current flow from positive to negative.

Current The movement of electrons is referred to as current flow. Some effects of current flow are heat, magnetism, and chemical change.

Cycle A complete sequence of wave pattern that recurs at regular intervals.

Density Compactness: The density of a magnetic field is increased by an increase of current flow in the magnet coil.

Dielectric An insulator: A material that does not permit the easy movement of electrons.

Differential amplifier The output voltage of a differential amplifier is an amplification of the difference between two input voltages.

Digital Method of sending information through an electrical circuit by switching the current on and off or by switching between a high and a low voltage.

Digital switch A device capable of only two outputs: high and low or on and off.

Diode A device that will permit electric current flow in one direction and blocks electric current flow in the opposite direction.

Direct current Electron flow in a circuit in one direction.

Driver A transistor in the output side of a computer that controls the electrical activators.

ECA Electronic Control Assembly: A term for an automotive computer.

EGO Exhaust Gas Oxygen sensor: Feeds electrical signals to the computer based on the amount of oxygen present in the engine exhaust, which is representative of the fuel/air ratio feeding the engine.

Electric current See Current.

Electrode Either of the two conducting elements where current enters or leaves certain electrical devices.

Electron theory A practice of tracing electric current flow from negative to positive.

Feedback The return of a small part of the energy from one point in an electrical system to a preceding point as from output to input.

Fixed resistor A resistor having a definite value that cannot be changed.

Frequency The number of cycles of a periodic function, usually in cycles per second.

Fuel injector In a gasoline engine, a solenoid valve that is spring-loaded shut. On electrical impulse from the computer, it permits fuel from a pressure line to spray into the intake manifold.

Gain Ratio between the input signal and the output signal of an amplifier.

Ground A common point in an electrical or electronic system. Usually considered 0 voltage.

Ground side switching Switch located between the electrical component being controlled and the ground side of the circuit.

Hall effect The voltage developed between opposite edges of a current-carrying metal strip when exposed to a magnetic field.

Hertz (Hz) A unit of frequency equal to one cycle per second.

IC Chip (Integrated Circuit) An IC is a complete electronic circuit containing all components and interconnecting conductors processed on a single chip of silicon.

Impedance The total opposition a circuit offers to the flow of alternating current. It includes resistance and reactance and is measured in ohms.

Induction The creation of a voltage in a conductor when it is passed through a magnetic field.

Input A signal put into an electrical or electronic device.

Insulator A material that, for all purposes, will not let electric current flow through it.

Interface A device that connects different sections of the computer together or connects the main computer to its external sensors and actuators.

Kilo Kilo = 1,000.

Knock sensor An electronic device capable of measuring vibration and converting the vibration to an electrical signal. Tuned to detect engine spark knock, its output will cause the computer to reduce the timing advance.

Light emitting diode (LED) A semiconductor that emits light when current is passed through it.

Linear variable differential transformer (LVDT) Used as the sensing device in some types of MAP sensors. Converts changes in engine manifold pressure to electrical signals that can be interpreted by the computer.

Liquid crystal display (LCD) Uses a liquid that has the ability to block or pass light, depending on the application of an electrical signal.

Magnet A device that has an attraction for iron.

Mass The quantity of matter in an object.

Mega (M) One million.

Micro (μ) One-millionth.

Microprocessor An I/C (integrated circuit) or a set of I/C's that can perform a variety of operations.

Milli (m) One one-thousandth.

Mixture control solenoid Operating on electrical signals from the computer to modify the carburetor fuel/air ratio.

Negative temperature coefficient (NTC) Description given to a substance in which the electrical resistance decreases as the temperature increases.

Nonvolatile RAM Random access memory that retains all of its properties (memory) when the power is turned off.

Ohm Unit of measure of electrical resistance. Amount of resistance in a circuit that will allow one ampere of current to flow with one volt supplied.

Ohm's law The relationship of voltage (E), resistance (R), and current flow (I) in a circuit: $I=E/R$. The current in a circuit is directly proportional to the applied voltage and inversely proportional to the resistance.

Open loop System operates on a fixed program, does not use system output as an input to the computer.

Output (1) An information signal going out of a system. (2) The connections where the information signal is delivered.

Parallel circuit A system that has more than one path (circuit) for current to flow.

Permanent magnet generator A generator that uses a permanent magnet to produce its magnetic field rather than an electromagnet.

Permeability The measure of how much better a material provides a path for a magnetic field than air does.

Photocell A light-sensitive device that changes variations in light intensity into corresponding voltage change.

Photoelectric The ability to produce voltage by exposure to light.

Polarity Having two opposite charges (positive and negative) determines the direction of current flow.

Polarize To establish polarity.

Port fuel injection (1) A system where the fuel is injected into the intake manifold right next to the intake valve. One injector is provided for each cylinder. (2) In a port system, the injectors are usually connected electrically, two or more in parallel.

Positive temperature coefficient Description given to a substance in which the electrical resistance increases as the temperature increases.

Potentiometer A variable resistance having two fixed terminals attached to opposite ends of a resistor and a third contact that moves up or down the resistor.

Pulse width The time that a fuel injector (or other output device) is "turned on." Usually measured in milliseconds.

RAM Random access memory: A form of computer memory that can be changed as well as read from.

Reactance Total opposition to the flow of an alternating current or pulsating direct current. Measured in ohms.

Readout The method the computer uses to display the processed information.

Reference voltage Regulated power supply for some system sensors that is provided by the computer.

Relay An electromagnetic switch. A small current in the control (coil) circuit can control a large current in the load (contact) circuit.

Reluctance The opposition of a material to magnetic lines of force.

Resistance Measured in ohms (Ω). The opposition in an electrical circuit or device to the flow of electrons (current).

Rheostat A variable resistor with one fixed and one movable contact.

ROM Read only memory: A form of computer memory that is permanent, cannot be changed.

Semiconductor A material that is neither a good conductor nor a good insulator. It can be either, depending on how it is connected in a circuit.

Sequenced indicators Indicators (usually lights) controlled by the computer. Will only light when the indicated system is out of range.

Sequential fuel injection A fuel injection system having an injector in the intake manifold for each cylinder. The injectors are operated individually in the engine firing order.

Series circuit A circuit in which components are connected end to end so the same amount of current flows through the entire circuit.

Signal arm (potentiometer) The movable part of a potentiometer.

Solenoid, normally closed An electrically operated valve that is spring-loaded closed.

Solenoid, normally open An electrically operated valve that is spring-loaded open.

Stepper motor A direct current motor that rotates its armature in steps (or specific amounts) in either direction. Operation is controlled by the computer, which operates specific windings in the motor.

Stoichiometric The air/fuel ratio needed for perfect combustion in an engine (approximately 14.7 to 1).

Telltale Warning lights to alert driver to system problems.

Thermistor A solid state semiconductor device whose electrical resistance changes with temperature. Thermistors have a negative temperature coefficient.

Throttle body The assembly in the air inlet system of a fuel injected engine that contains the throttle valve.

Transistor A solid state semiconductor device with three or more connections. Used to amplify and control electrical signals in electronic circuits.

Vacuum tube fluorescent Makes use of a phosphor material that emits light when bombarded by electrons, in the same manner as a TV picture tube.

Variable reluctance (magnetic) A means for increasing or decreasing the strength of a magnetic field.

Voltage A measure of electrical force or pressure. It takes one volt to move one ampere through one ohm of resistance in an electrical circuit.

Voltage drop The amount of electrical pressure (voltage) used up in a circuit or part of a circuit when the circuit is in operation.

Voltage regulator A device to regulate or maintain electrical pressure (voltage).

Voltage spike Surges of voltage above the normal regulated level.

Volume (1) space occupied. (2) The intensity of a sound.

Volumetric efficiency The pumping efficiency of an engine as air is drawn into the cylinders.

Wheatstone bridge (strain gauge) An instrument for measuring differences in resistance.

Index